BEYOND PEPPER SPRAY

The Complete Guide to Chemical Agents, Delivery Systems, and Protective Masks

Michael E. Conti

Paladin Press • Boulder, Colorado

Order and simplification are the first steps toward the mastery of survival—the actual enemy is the unknown.
—Thomas Mann, *The Magic Mountain*

Beyond Pepper Spray:
The Complete Guide to Chemical Agents, Delivery Systems, and Protective Masks
by Michael E. Conti

Copyright © 2002 by Michael E. Conti

ISBN 1-58160-291-X
Printed in the United States of America

Published by Paladin Press, a division of
Paladin Enterprises, Inc.
Gunbarrel Tech Center
7077 Winchester Circle
Boulder, Colorado 80301 USA
+1.303.443.7250

Direct inquiries and/or orders to the above address.

PALADIN, PALADIN PRESS, and the "horse head" design
are trademarks belonging to Paladin Enterprises and
registered in United States Patent and Trademark Office.

Visit our Web site at www.paladin-press.com

Contents

SECTION II

Aerosol Subject Restraint Sprays
Smoke and Chemical Agent Weapon Devices
Chemical Agent Protective Mask
Individual Operation Skills
Unit Operation Skills

This book is dedicated to the memory of Col. Rex Applegate, a good soldier who stayed on mission to the very end.
Col. Rex Applegate, 1914–1998.

Warning

The purpose of this book is to present an overview of modern chemical agents that are not primarily lethal and of their delivery systems to the trained police officer, member of the military, or security professional. It is in no way meant to replace or contradict any department's or agency's current policies or procedures.

The justified use and employment of any type of weapon,[1] regardless of its intended level of force, is a subject that must be addressed by individual departments or agencies. Federal law; state law; and departmental rules, regulations, policies, and procedures must all be satisfied.

Today, in the United States the use of deadly force by law enforcement officers is generally permissible only as a last resort, when in the reasonable and considered opinion of the police officers there is an imminent threat of death or serious bodily injury to them or others.

Lower levels of force, including the employment of chemical agents and specialty impact munitions, may be used much earlier in a confrontational situation. Depending on the facts of the situation, these agents or munitions may be used in conjunction with, or instead of, other less-lethal force options (such as batons or empty-hand techniques) while effecting an arrest or attempting to control an out-of-control individual or group.

It is imperative that each officer operating in the field knows and understands both the meaning and the intent of all applicable standards regarding the use of any level-of-force option available to him. It is also imperative that he be able to articulate his actions clearly and reasonably, *especially* when those actions involve any level of force.

As law enforcement, military, and security professionals, we must also be familiar with various federal statutes and court rulings that have a direct impact on what actions we may take and how they may affect our liability exposure while performing our duties.

Your department, unit, or agency is responsible for your training in this area. Once he is properly trained, the onus of responsibility for adhering to this training is then squarely on the individual officer.

Neither the author nor the publisher is responsible for the use or misuse of any information contained in this book. It is presented for *academic study only*.

NOTE

1. *Webster's New World Dictionary* defines *weapon* as (1) any instrument used for fighting; (2) any means of attack and defense.

Foreword

We haven't had a comprehensive book on the use of chemical munitions since Thomas Swearengen's classic book *Tear Gas Munitions* or since Rex Applegate gave us the book *Riot Control*, which many consider to be the definitive work of its type. Colonel Applegate was in the process of rewriting his book to bring it up to date with the many technical advancements in the field when he passed away. It is unfortunate that the Colonel never finished that project.

The landscape of the law enforcement chemical munitions world has changed drastically since both Applegate and Swearengen made their contributions. We must now be more concerned about litigation resulting from our use of chemicals as an aid to resolving violent confrontations than ever before. Occupational safety and environmental issues that previous authors never imagined now concern even the smallest agency that seeks the tremendous advantages of using these tools. It is hard to overestimate the effect of the development of oleoresin capsicum products on tactics, dissemination methods, and even the public perception of the use of chemicals in law enforcement.

Gilbert DuVernay, director of the Smith & Wesson Academy, Springfield, Massachusetts.

Although many of today's products would be very familiar to an officer of the 1960s, the 1950s, or even the 1940s, many others bear little resemblance to anything used by our predecessors in this challenging field. By the same token, many of the products that they accepted were eliminated when experience showed them to be difficult or even excessively dangerous to use. The situation is vastly changed since those great contributions of the past, and now the need again exists for a book that puts it all in perspective.

Mike's book fills that need. It is a well-researched modern guide for today's peace officer in the selection, use, cleanup, and disposal of law enforcement chemical agents. It will be an often-used resource for anyone with responsibility, or an interest, in this field.

—Gilbert DuVernay
2001

Author's Notes

For the record, I have no financial interest in any chemical agent or protective respirator manufacturing company that produces any items illustrated in this book or the distribution companies that market them.

The descriptions of the products, as well as opinions expressed, are my own, based on my training, experience, and personal beliefs.

The only obligation I operate under is to provide members of the law enforcement community with the most accurate and comprehensive information that I can gather.

LESS-LETHAL FORCE DEFINED

Less-lethal force can be defined as a level of force *less likely* to inflict a wound or injury that would result in a fatality. This term is primarily used to describe an application of force that involves weapons other than firearms loaded with standard penetrating projectile-type ammunition.

The use of *less-lethal* is actually more precise than the often used phrase *less than lethal* because this second term may be taken to imply that a weapon or technique is patently *nonlethal*.

Any application of force can, under certain conditions or circumstances, produce a lethal result.

Therefore, *less lethal* will be used throughout this work. The adoption of this more specific term is also highly recommended for department and agency use.

Acknowledgments

I would like to acknowledge and thank all those who helped make this book possible.

First, I want to express my gratitude to the folks at Paladin Press for their support of my efforts. Jon Ford, Donna DuVall, Barb Beasley, and everyone else at Paladin have been a pleasure to do business with during the production process of both this book and the first in the series, *In the Line of Fire: A Working Cop's Guide to Pistolcraft*. Jon's patience was especially appreciated because my full-time work schedule didn't exactly allow me to finish this book in "about six months," as I first predicted. It actually took a little closer to three years! Never once did Jon make me feel pressured or remind me of my initial arrogance regarding the amount of time it would take me to research, write, photograph, illustrate, and ship him a completed manuscript. I consider myself lucky to have been initially steered to Jon and Paladin by Denny Hansen, editor of *SWAT* magazine.

Al Pereira, who produced the majority of photographs used in the first book and many of those in this work, is a class act as well as a good friend. When he wasn't behind the camera taking the shots himself, he was helping me take them by advising on lighting, setup, and composition—actually, he did pretty much everything but press the button. His patience and goodwill are much appreciated.

I would also like to thank Mark Bollhardt, Dave Young, Kevin Williams, and Dawn Flynn, formally of MACE Security International; Jim Cavanaugh, formerly of Defense Technology Corporation of America; Kevin and Jill Dallett of Aerko International; Gigi Doyle of Reliapon Police Products, Inc.; Brian Carbone of New England Public Safety Supply, Inc.; Todd Resnick of Fume Free, Inc.; Herb Schribner and Mark Schramm of Guardian Protective Devices, Inc.; Jim Perry of ALPEC Team, Inc.; and Ed Ferguson of Fox Labs, Inc. All were very generous with their time and knowledge and provided me with samples of many of their latest chemical agent products for testing and evaluation.

Jeff Brown, of Law Enforcement Targets, also contributed materials to this book, as he did with the first. Louie F. Leitão, of Jane's Information Group, was of immeasurable assistance by allowing me access to the incredible stores of information compiled in the Jane's reference books.

A special thanks to Howard Sarris, affectionately known as "the Gas Peddler," who let me pick his brain and shared colorful stories from his many years of working in the chemical agent weapon industry.

A great debt is also owed to Bert DuVernay, director of the venerable Smith & Wesson Academy in Springfield, Massachusetts. Bert, a true professional with an inquisitive mind, tolerant nature, easygoing manner, and an always busy schedule, was good enough not only to review this work prior to publication, but also to write the foreword for it. I knew Bert was the man I wanted to write the foreword after he and I spent about three hours on a warm spring day digging through buckets of old chemical agent munitions. At one point both our respective eyes lit up when a pristine DM pyrotechnic projectile was pulled from one of the poison-placarded gray buckets. During the mild argument that ensued over who would add the perfectly preserved, shiny green munition to his collection, we both came to the conclusion that we were probably the only two people in North America who would consider a "Vomiting Agent" munition as a highly desirable collectible. Despite the fact that he walked away with that particular prize, I highly recommend any classes Bert and his staff offer: they have always proven to be well worth the time and expenditure of training funds—and that is high praise indeed, coming from the underfunded world of the police industry.

I am also very grateful for the time and assistance provided to me by Paul Damery, Tim Donnelly, Marty Driggs, Roger Ford, Rich Lane, Patrick McAdam, and Tom Robbins. Some people are just always there when you need them, no matter what.

Thanks also to all the other officers, soldiers, and operators who allowed me to photograph them or pick their brains while working over the years, including the late, great, Col. Rex Applegate. Colonel Applegate's passing came too soon, but, as a testament to the type of extraordinary soldier that he was, he suffered the stroke that resulted in his death seven days later while on mission *at age 83*, still trying to save lives by sharing the wealth of his knowledge and experience. Neither he nor his work will be forgotten.

On the home front, to my wife, Kathy, I extend my heartfelt gratitude for all the time and effort she has put into supporting my efforts. Always willing to listen and with a sharp editing eye, she has been a tremendous help as well as a great wife and mother to our two wonderful children, Katie and Nick. Katie, then 4 years old, also pitched in by posing for the first photo that appears in Chapter 1. Now 7, she's writing her own books on her own computer. How times have changed. . . .

To my parents, Margie and Jim, thank you for teaching me that you must never stop trying in this world and that a single step, day after day, truly will get you to the other side of the mountain.

Finally, to my extended family, the members of the Massachusetts State Police and law enforcement in general, I wish to express my gratitude for all the things I have learned and continue to learn from them. Honesty, loyalty, tenacity, and perseverance— these are the traits I see most often demonstrated by my fellow officers. And these traits continue to flourish even during these times when it is seemingly more popular in America to doubt and question the word of a police officer than that of a convicted felon.

So be it. All of the warm and fuzzy dogma aside, in the long run it will be as it has always been, for we live in a society that ardently demands results from the police, yet is vehemently opposed to the idea of *being* policed.

What that leaves those of us in law enforcement with are two things—our core desire to do the right thing and *each other*. Both must be preserved at all costs.

Introduction

The use of chemical agent weapons in the United States and abroad has seen a decided upturn over the past decade, especially in the use and employment of oleoresin capsicum (OC), or so-called pepper spray. Whether dispensed by aerosol spray, hand-thrown canisters, or shotgun fire, OC has radically changed the way law enforcement perceives the role and viability of these versatile tools.

Unfortunately, there is a great deal of misinformation currently circulating about both the use and employment of OC, as well as other chemical agents. The purpose of this work, therefore, is to identify the various chemical agent munitions available to law enforcement, OC included, and explore their uses—past and present—in a realistic and commonsensical manner. Such equipment as gas launchers, foggers, and respirators is also addressed.

I attempted to keep two things in mind while working on this book. First and foremost is the KISS principle (Keep It Simple, Stupid), which I find indispensable both while working and when off duty. The second thing is the pursuit of common sense—and I do not use the word *pursuit* lightly here. Common sense, as most of us know, has a unique way of fleeing the premises whenever certain subjects are brought up. Very often, anything directly involved with, or even remotely related to, the police industry qualifies.

Too many people trying to push too many products to an industry with decidedly high-quality tastes and low-bid budgets can induce an effect decidedly lacking in common sense. Not to mention a lot of well-intentioned administrators who latch on to one product or training theory over another for any number of reasons, not least because it has been presented to them in a way that strikes their fancy.

Let me assure you now that this criterion has *not* been applied to this book.

Everything presented on these pages is as technically accurate and reality-based as I can make it. All tactics or techniques discussed have been practiced and employed in the field by me or others about whom I have firsthand knowledge. Personal experiences relevant to the specific materiel covered are also included. Some of these experiences, far from being "war stories" presented to glorify an officer, may relate embarrassing or undesirable situations. We have all had them. It is my belief that these experiences may be among the most beneficial we can have because they keep us thinking and ready to react to the unexpected.

It is how we learn. That is, of course, as long as we do not put the mistakes and their valuable lessons out of our minds to avoid embarrassment or bolster our ego.

Naturally we all desire to succeed in whatever situation we find ourselves, the entire purpose of the exercise being to go home at the end of the shift and prevail in any attendant civil litigation. So while failure is never an option, we must realize that mistakes and snafus are inevitable. Those related here are presented in the hope that others may avoid them. I would like to make it clear, however, that while I tend to use the occasional error to illustrate potential problems to be aware of, the vast majority of missions undertaken by the members of U.S. law enforcement agencies are successful. This is overwhelmingly due to the incredible (and mostly unrecognized) amount of hard work countless individuals and units perform, day in and out, all across our nation.

This book is formatted to allow easy access to the information presented. The second section of the book contains some basic training guidelines for consideration. The exercises presented are reality based, field useful, and simple to perform. Many also address real-world training concerns that are often overlooked.

Technical statistics and chemical formulas are included at the end of the book for those who either desire or require them. Also included are a list of suggested additional reading, a list of contacts for equipment and training, and a glossary of terms used in this book.

I hope that the material is useful to the reader. I also hope that the police officer, military personnel, or security specialist who reads the material does so with a critical, questioning eye—because no one has all the answers.

Like most of us, I am occasionally not even so sure of the questions. But what I am sure of is this: it is our duty to one another as law enforcement professionals to continually strive toward improving the training and safety of all of us in the law enforcement community. That is why, even though I have done my best to bring you the most accurate information available, I respectfully request that you let me know if you find my effort here lacking or inaccurate in any way, or if you have developed or are aware of more efficient or reliable equipment or techniques than those shown on these pages. I would sincerely welcome any suggestions or comments, positive or otherwise.

I can also assure you that I will do my best to include any new ideas, tactics, or techniques that prove viable in later editions of this book and that credit will be given to the contributing individual, department, or agency. I can be reached either through Paladin Press or at the address listed below.

Thank you in advance for your consideration in this matter. I hope you enjoy the book.

Michael E. Conti
c/o Saber Group, Inc.
268 Main Street, No. 138
North Reading, MA 01864
E-mail: mconti@sabergroup.com
Web site: www.sabergroup.com

SECTION I

Safety

WHAT IS A CHEMICAL WEAPON?

The term *chemical weapon* may be confusing because it is sometimes used as a generic description of any type of chemical agent intended for use against human beings. The term *chemical weapons* and the specific types of agents they refer to, however, have been fairly well defined since the mid-1950s when the military began to use the agent orthochlorobenzalmalononitrile, or CS. Prior to this, *chemical weapon* referred to specific classes of chemical warfare agents (toxins and lethal and incapacitating agents) that could cause death or serious injury in normal field concentrations.

Because of its ability to affect the mucous membranes at extraordinarily low concentrations while providing an extremely high safety ratio that makes it almost impossible to create a dangerous dose, CS was given the (then) new designation of "riot control agent."

In the years since, various riot control agents have been adopted and used by both the military and civilian police industries in the United States and abroad, and the generic use of the term *chemical weapon*, though technically incorrect, has become more common when referring to them.

This is understandable since chemicals are indeed used in both, and the common definition of the word *weapon* is "any instrument used for fighting or for means of attack and defense."[1]

So, in an effort to avoid any confusion between weapons that employ true chemical warfare agents

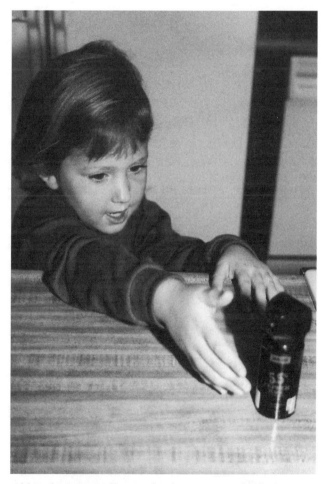

Although not generally considered to constitute deadly force, chemical agent weapons are very dangerous in small or untrained hands. Just like any firearm, they must be properly secured when not in use. Also like the firearm, the more accessible the chemical agent weapon, the more liable the owner is for any use or misuse of it.

and those that employ riot control agents, I use chemical weapon (CW) and chemical agent weapon (CAW) throughout this book and would respectfully suggest them for general use.

Chemical Weapon: This is any ammunition or equipment that disperses any chemical substances—whether gaseous, liquid, or solid, classified as a toxin, lethal, or incapacitating agent—that can cause death or serious injury in normal field concentrations.

Chemical Agent Weapon: This is any ammunition or equipment that disperses any chemical substances—whether gaseous, liquid, or solid, classified as a riot control agent—that will generally not cause death or serious injury in normal field concentrations.

In a final attempt to avoid confusion, I believe it is prudent to employ a separate descriptive term for chemicals designed to produce smoke that does not contain CW or CAW classified ingredients. For even though certain types of smoke screen may be poisonous in extremely high concentrations, smoke ammunition is not intended to cause death or serious injury or to produce the direct physiological effects of riot control agents.

Smoke Ammunition: This is any ammunition or equipment that produces obscuring smoke intended for use as a signaling tool or as a means of providing concealment for movement of personnel.

A WEAPON IS A WEAPON

Safety is one of the most important considerations in any type of training. When weapons are involved, the level of safety awareness must be increased in direct proportion to the potential damage the weapon is capable of inflicting.

Most people adhere to this philosophy when dealing with firearms, but for one reason or another, they either dismiss it or pay it little heed when dealing with weapons considered to be less lethal than the issue pistol, rifle, shotgun, or other firearm. The problem with this attitude is the mind-set it generates.

The instructor is responsible for setting—and maintaining—the standard of safety during any training program.

When involved in any type of training, the persons being trained will generally key on the instructor, following his lead and example. If the use of a less-lethal weapon is the object of the training exercise, and the instructor disregards the basic tenants of safety or downplays them, then this attitude will more than likely be reflected in the students—often both while they are being trained and later when they are on the street employing these same weapons.

Besides creating an unsafe and unprofessional atmosphere, there is one other major flaw in this type of thinking—it does not recognize or take into account the fact that weapons intended to be less lethal can very easily inflict or result in lethal injury should they be used improperly (or properly, should the circumstances require it). This is true regardless of the weapon. Batons, CAWs, launched projectiles, even empty-hand techniques can all cause serious injury or death—even though they are often considered as less than lethal, they are in no way nonlethal.

That is why in addition to examining the proper methods of CAW employment, we will look at how the various CAWs and their delivery systems can cause more than their normally desired effects. This aspect needs to be addressed because causing the unintentional death of a subject while employing a supposed less-lethal tool or technique is unacceptable, especially if that subject does not present an immediate threat of death or serious bodily injury.

The limitations of these weapons will also be examined. Just as unacceptable is the idea of an officer's employing a less-lethal classified weapon against a subject presenting a lethal threat when it puts the officer at a distinct disadvantage and in true peril.

The bottom line regarding the safe use and employment of less-lethal classified weapons is this: A weapon is a weapon, and safe weapon handling is everyone's responsibility, all of the time!

TEN COMMONSENSE SAFETY RULES FOR HANDLING LESS-LETHAL WEAPONS

1. Treat every weapon with the same respect given a loaded firearm.
2. Do not unnecessarily handle or display any type of weapon unless you are intending to use it.
3. Know the realistic capabilities and limitations of any weapon you carry.
4. Always train with your equipment as you will carry it in the field.
5. Be prepared to immediately discard your less-lethal weapon and use your firearm should the situation escalate to one of deadly force.
6. Secure *all* weapons when not in use—keep them away from small or untrained hands!
7. Regularly inspect canisters, nozzles, seals, spoons, and pins for deformations or signs of wear.
8. Use only proper, issued munitions and chemical agents.
9. Avoid any kind of drug or alcoholic beverages when handling weapons of any type.
10. A weapon is a weapon.

MATERIAL SAFETY DATA SHEETS

Material safety data sheets (MSDSs) are forms prepared by manufacturers of hazardous materials. The preparation of these forms is required to comply with Occupational Safety and Health Administration (OSHA) Hazard Communication Standard (29CFR 1910. 1200).

The forms normally list all pertinent information about the material's origin, identification, chemical properties, handling, and emergency procedures should something happen to it during storage or shipment. The forms also identify any physical or health hazards that the material may pose.

Generally provided to transportation companies for use by their personnel, the forms also contain a lot of relevant information that may be useful to departmental armorers, chemical agent officers (CAOs), instructors, and any other officers who may employ or otherwise come into contact with the agents.

Some departments actually issue a copy of the appropriate MSDSs to each officer equipped with any type of CAW. This form, kept handy in the cruiser, can then be provided to medical personnel should anyone exposed to the agent suffer an adverse reaction to it and require treatment.

Be advised, however, that some MSDSs may not list all ingredients in a munition, especially in the case of aerosol subject restraint sprays. This is because federal rules allow manufacturers to conceal ingredients as "trade secrets." If this is the case with the agent you have been issued or are carrying, then it is strongly suggested that you or someone from your department contact the manufacturer and request a full disclosure of ingredients. If you are unable to secure this information, then serious consideration should be given to adopting a different CAW for use. Otherwise, you or your people will be placed in the unenviable position of carrying and employing chemicals you can't identify. Should there be a medical emergency, this ignorance could prove costly.

Regardless of the type or quantity of chemical agents your department maintains or issues, a complete MSDS for each should be kept on file. Your department armorer or CAO should also be familiar with the contents of all the MSDSs kept on file and be prepared to produce them should the need arise. This may occur as a result of injury or death directly caused by—or indirectly related to—exposure to any chemical agents your personnel use.

In this event, it obviously would be far more professional for your CAO to have a current MSDS for any agent used, as opposed to having a plaintiff's attorney present one to a jury.

Keeping in mind that chemical agents are classified as hazardous materials—including OC sprays, depending on the carrier agent used—the effect on a jury could be overwhelming, especially if the officer responsible for issuing, training with, and using the chemical simply stares at the MSDS from the witness stand with an expression similar to that of a hog staring at a wristwatch.[2]

A sample MSDS is found in Appendix C.

NOTES

1. *Webster's New World Dictionary*.
2. Carrier agents such as trichlorotrifluoroethane, though considered low in toxicity, are still classified as hazardous materials.

Chemical Weapons and Chemical Agent Weapons: An Overview

Absent a determined program of action . . . Americans have every reason to anticipate acts of nuclear, chemical, or biological terrorism.
— Sen. Richard Lugar (Republican-Indiana)
31 October 1995

One hundred grams of anthrax properly dispersed downwind over Washington, D.C., for example, could kill between 150,000 and three million people in the surrounding areas. . . . The ever faster pace of technological advance . . . has made the know-how for creating, manufacturing, and dispersing these agents of mass destruction widely available through open literature. . . . A dozen countries, including all states named by the State Department as sponsors of terrorism . . . have developed bio-warfare capability.
— Dr. Tara O'Toole, deputy director of the Johns Hopkins University Center for Civilian Biodefense Studies, 22 August 2000

It is believed that one of the first uses of CWs occurred more than 4,000 years ago when smoke screens, incendiary devices, and "toxic fumes" were used in ancient India.[1]

Approximately 2,000 years ago, Chinese warriors discovered that an enemy could be effectively blinded if finely ground pepper was thrown into his eyes. (Ironically, some two millenniums later, modern man would rediscover the viability of this simple CAW with the development and use of OC, which is, for all intents and purposes, the same thing.)[2]

In between, many other types of CWs and CAWs have been developed, tested, and employed. Not surprisingly, the majority of these were originally produced for use against armed combatants in military conflicts.

Riot control agents, developed (and generally considered) to be less lethal, were eventually drafted for use by the military and the police in the United States and many other countries. For instance, chloroacetophenone (CN), developed just before World War I, was used to train and prepare soldiers to deal with the deadly phosgene and chlorine gases employed on the battlefield. After the conclusion of hostilities, the men of the American Expeditionary Forces returned to the United States, and many of them gravitated into careers in law enforcement.

As the years following the end of the Great War progressed so did civil unrest, labor strikes, and organized crime. It was in this environment that the idea of using the mild CN agent against rioters and other criminal elements became very attractive to nation's law enforcement agencies. The use of CN and other agents that have come to be classified as riot control agents has continued since that time both in this country and throughout the world. Today, they are still the only type of chemical agents sanctioned for use against civilians in most countries, including the United States.

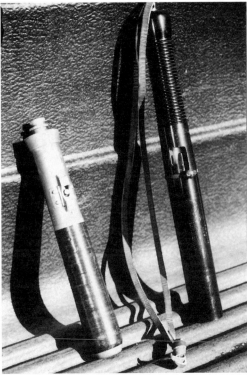

Above Left: A "tear gas test" being conducted at the army base in South Boston in April 1925. This demonstration was reportedly made to prove the defensive powers of the tear gas against the threatened attacks of bandits, thugs, and the like. Federal Laboratories' "Gas Billy" (shown later) was also demonstrated during this test. (Original photo from the author's collection, photographer unknown.)

Above Right: The "Gas Billy" was originally designed for use by bank messengers so that they would "be able to withstand the attack of holdup men" by "pointing the bill at the bandit" and "letting the tear gas do its dirty work." Later used by law enforcement officers, the Gas Billy fired a 12-gauge round loaded with (usually) CN when the spring-loaded striker in the handle was released. The agent would be expelled from the business end of the club. The device could also be used as an impact weapon. Shown in the photo are two different models. The Lake Erie Chemical Company P-31 Police Gas Billy on the right used a simple spring-loaded striker that was retracted manually and released. Federal Laboratories' Model M-14 on the left was fancier, using a leather-bound barrel and push-button striker release. (These Gas Billies are from the author's collection.)

Gas Billy disassembled for loading. A 12-gauge gas round would be loaded and the unit reassembled for use. (Gas Billy from the author's collection.)

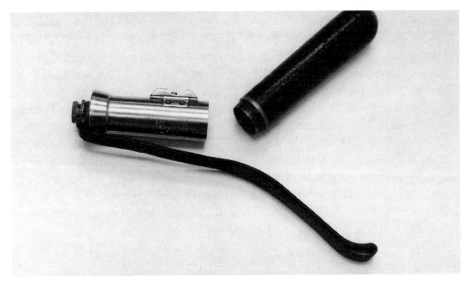

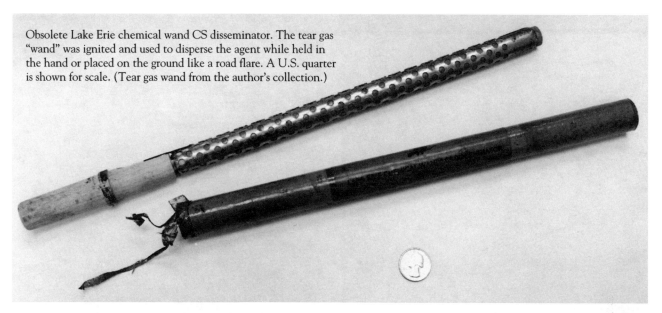

Obsolete Lake Erie chemical wand CS disseminator. The tear gas "wand" was ignited and used to disperse the agent while held in the hand or placed on the ground like a road flare. A U.S. quarter is shown for scale. (Tear gas wand from the author's collection.)

Although toxins (as well as the biological organisms that produce them), lethal agents, and incapacitating agents are classified as chemical warfare agents and therefore fall outside the scope of this book, I believe a brief description of each is warranted.

Unfortunately, these particular types of weapons are sought by—and available to—terrorist and extremist groups, as well as many developing nations, some with unstable governments.

As events in the Persian Gulf and Japan have shown, the possibility of such production and deployment of these agents is anything but far-fetched. It has been and continues to be done.

And as terrifying as the thought may be, the world is still populated with the following:

- Dictators such as Saddam Hussein, who is known to have used chemical weapons and is also strongly believed to still be in possession of great stockpiles of both chemical and biological weapons
- Terrorist groups such as the German Red Army Faction, which in a 1984 Paris raid was discovered to have not only documents indicating a solid working knowledge of lethal biological agents, but also a bathtub full of flasks containing *Clostridium botulinum* (which produces botulism)
- Cults such as the Japanese Aum Shinri Kyo, or "the Supreme Truth," which used chemical agents in several terrorist attacks in 1994 and

1995, culminating in the Tokyo subway attack of 20 March 1995 when members released sarin gas, leaving 12 people dead and more than 5,000 injured

The above are just a few examples of what's out there. Combining the availability of these "poor nation's" weapons of mass destruction with the continuing increase of terrorist attacks both at home and abroad, I believe it is in our best interest to be aware of the existence of these weapons, as well as having a basic understanding of what they are and how they work.

For when the unthinkable occurs again, we in the military, law enforcement, and security fields will be among the first called on to respond.

BIOLOGICAL AGENTS

Biological agents may be placed in two basic groups: one comprising living pathogenic microorganisms, which are mainly such live organic germs as anthrax (*Bacillus anthrax*); and the other consisting of toxins, which are the byproducts of living organisms. One of these effectively natural poisons, botulinum toxin, is produced when the microorganism *Clostridium botulinum* is grown. In layman's terms, botulinum toxin is referred to simply as botulism—one of the deadliest biological substances known to man.

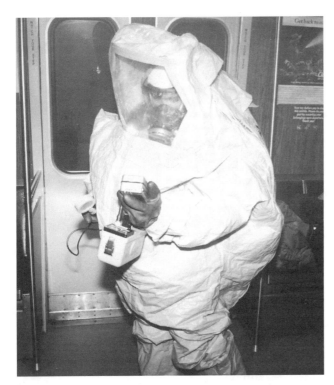 

Members of the Boston Fire Department Emergency Response Unit wearing full hazardous materials (HAZMAT) suits check for contaminants (left) and survivors (right) on subway during terrorist-incident preparation training exercise.

Role players portraying victims who have been incapacitated by a nerve agent. Training exercises similar to this have been conducted across the United States during the past few years.

Large-scale terrorist-incident scenarios give the various first-response elements a chance to work together, evaluate options, and prepare for the realities of a CW attack. Here, members of federal, state, local, and transit law enforcement agencies participate in a debriefing along with members of the military, fire department, and Federal Emergency Management Agency.

The use of biological weapons, sometimes referred to as germ warfare, is by no means a modern development. Germ warfare—the intentional use of diseases to affect an adversary's military force, population, crops, or livestock—has long been considered an effective combat weapon. There are numerous recorded instances of its use throughout the centuries.

In the 6th century B.C., the Assyrians reportedly poisoned enemy wells with rye ergot. Ergot is a fungus blight that forms hallucinogenic drugs in bread. When consumed, the victims of ergot can be afflicted with a disorder called St. Anthony's fire. Paranoia, hallucinations, twitches and spasms, cardiovascular trouble, and stillborn children are symptoms of this disorder, as is a weakening of the immune system.

In A.D. 1346, the Tatars successfully used biological weapons in the form of plague-infected bodies that were catapulted over the walls of the besieged city of Kaffa (now Feodosiya, Ukraine). This tactic not only broke the siege, causing the city to capitulate, but also has been cited by some historians for having started the bubonic plague epidemic that swept across Europe over the following five years, killing more than 25 million people.

The Romans fouled their enemies' water supplies by throwing animal corpses into wells. The British gave smallpox-tainted blankets to Indians loyal to the French during the French and Indian War. Confederate forces also used dead animal carcasses during the U.S. Civil War to contaminate Union water supplies. After World War I, Japan dropped flea-infested debris over cities in China, resulting in a bubonic plague epidemic. The British, Americans, Canadians, and Soviets also developed and experimented heavily with biological agents during the years after World War I.

The development and stockpiling of biological weapons throughout the world continued at a feverish pace until about 1969, when President Richard Nixon drastically changed the U.S. policy on biological warfare. His decision to terminate all research on biological warfare was the catalyst for the establishment of the Biological and Toxin Weapons Convention (BWC), which the United States and many other countries signed. The BWC stipulates that no member nation will, under any circumstances, "develop, produce, stockpile, or otherwise acquire or retain any biological weapons."

It is enlightening to consider that two of the member nations signing this unenforceable covenant were Iraq and the former Soviet Union—both of which have since admitted to violating it. Many other nations never even bothered to sign it, and its future is unsure as of this writing.

Regardless of the final disposition of the BWC, the cold reality of history provides perhaps the clearest indication of the value of such treaties. A similar covenant, arising from the Hague Conference of 1899, saw several nations (including Germany) pledge never to use suffocating or poisonous gases.[3] This agreement, which created a feeling of security throughout the world about the use of these munitions, was violated by Germany on 22 April 1915 during a surprise attack on French forces in the northern part of the Ypres Salient.

Perhaps the most disturbing aspect of the use of organisms, viruses, toxins, and biological agents as weapons is the ease with which they can be produced and delivered.

The more commonly known agents such as *Bacillus anthrax* (anthrax), botulinum toxin, *Yersinia pestis* (plague), ricin, staphylococcal enterotoxin B, and Venezuelan equine encephalitis virus (VEE) can be dispersed in aerosols with particle sizes of from 1 to 5 micrometers (microns). Under certain weather conditions, these agents can remain suspended in the atmosphere for hours. Once inhaled, the minute deadly particles will penetrate into the distal bronchioles and terminal alveoli of victims, resulting in death, panic, and horror as terrifyingly real as any experienced in the annals of history.[4]

TOXINS

As noted above, toxins are effectively natural poisons produced by living organisms. These poisons may be produced from animal or vegetable cells or synthesized in the laboratory. When these toxins are

inhaled or swallowed by or injected into human beings or animals, they cause illness or death.

Toxins are classed as chemical warfare agents if they are used for military purposes. However, they are most commonly grouped in with biological weapons and are in fact covered by the Biological and Toxin Weapons Convention, as well as the Chemical Weapons Convention.

LETHAL CHEMICAL AGENTS

Lethal chemical agents are those that may be used effectively in normal field concentrations to cause death. Among these are such choking agents as phosgene (CG) and diphosgene (DP), which attack the respiratory tract. Used extensively during World

German soldier and mount during World War I, circa 1918. The soldier is wearing a *Lederschutzmaske* (leather protective mask). Leather and other materials were used because of a shortage of rubber. (Photo from the author's collection, photographer unknown.)

War I, these agents—smelling like new-mown grass or hay—cause the victim to choke to death as the membranes in the lungs swell and the air sacs fill with watery fluid. The victim may then succumb to anoxia (oxygen deficiency), effectively "drowning" on dry land. Though originally developed in the mid-1850s, the first effective wartime use of these types of agents occurred in 1915.[5] The attack was initiated by the Germans against raw French colonial troops immediately following a heavy bombardment with conventional munitions.[6] The agent, first observed as "thick yellow smoke" flowing out of German trenches, had a devastating and deadly effect.[7]

Nerve agents such as tabun (GA), sarin (GB), soman (GD), and VX[8] attack the skeletal muscles, parasympathetic end organs, and the central nervous system. Despite being commonly referred to as "nerve gas," nerve agents are actually clear and colorless volatile liquids that evaporate, just as water and gasoline do, producing deadly, often invisible, vapors.

Nerve agents were first developed by a German research chemist, Dr. Gerhard Schrader, in 1936. Schrader, rather than trying to develop a weapon of war, was actually trying to improve insecticides when he synthesized the complex organophosphorous ester, later called tabun. GB and GD were developed prior to 1944, and the three are generally classified as G-agents. V-agent nerve gases (e.g., VX, VE, VM) were introduced by the British in 1948.[9]

These agents, either odorless or giving off a faintly fruity aroma, are quickly absorbed through the skin or eyes. Victims may exhibit a variety of unpleasant symptoms that include difficulty in breathing, drooling, sweating, nausea, vomiting, cramps, involuntary twitching, jerking, headache, confusion, coma, and convulsions. Death generally occurs very rapidly.

Blood agents—such as hydrogen cyanide (AC),[10] cyanogen chloride (CK), and arsine (SA)—are also classified as lethal agents. They are primarily absorbed into the body by breathing. They affect an enzyme, cytochrome oxidase, preventing the normal utilization of oxygen by the cells, causing rapid damage to body tissues. These agents, which produce aromas ranging from bitter almonds to a mild garlic-like odor, may be immediately lethal or simply induce light recoverable symptoms, depending on concentration, field conditions, and exposure time.

INCAPACITATING AGENTS

Incapacitating agents produce physiological or mental effects, which may last for hours or days. Their purpose, as the name implies, is to incapacitate combatants without killing them. One accepted rationale for their use is that an incapacitated soldier will not only be ineffective, but will also require help from others not affected, thereby taking more soldiers away from their normal duties and functions.

Agents such as 3-quinuclidinylbenzilate (BZ) disturb the higher integrative functions of memory, problem solving, attention, and comprehension. Other like agents are designed to cause excessive nervous activity, overloading the brain with too much stimulation. Other symptoms such as disturbed body awareness, disorientation, and vivid dreaming may also be experienced.

Complete recovery from the effects of these agents is usually possible without medical treatment.

Blister agents, though not actually classified as incapacitating agents, are also used primarily to cause casualties. They are used, as well, to deny terrain to the enemy, impede troop movements, and prevent materials and facilities from being used. Agents such as lewisite[11] (L), Levinstein mustard (HD or H[12]), and phenyldichloroarsine (PD), all fall under the heading of blister agents and cause similar effects.

The mustard agents (first used by the Germans against the British on 12 July 1917) act as cell irritants first and then as cell poisons, affecting all internal and external tissue surfaces contacted. The agents generally irritate the eyes, nose, and throat, and may bring on bronchial pneumonia. Vomiting, fever, skin lesions, severe diarrhea, and blistering may also occur. Lowered resistance to infections after exposure is also common. Mustard agents normally cause no immediate symptoms after first contact; a latent period, or delay of between 2 to 24 hours, may occur before the effects are noticeable.

In his report to the War Department, illustrating just how dangerously effective these incapacitating agents are, General John Pershing made these observations regarding the first mustard agent attack:

Following with regard to the new gas called "Mustard Gas" used by the Germans. . . . Since July 18 the British have suffered 20,000 casualties from this gas alone. Five percent have been fatal, 14 percent have been serious, while the balance have been mild. This so-called gas is really a liquid carried in shells which scatter the liquid on the ground. There is a slight odor of mustard of which men nearby are sensible. One can remain in this odor for five minutes or so without suffering any harm. The action is not noticeable for four or five hours after exposure. Then the soldier feels a burning of the eyes and a dryness and parched condition of the skin. He then becomes blind and the respiratory passages develop a condition similar to that in diphtheria. There are usually exterior burns on the exposed parts of the skin, hands, back of neck, and tender parts, as gas attacks the skin regardless of clothing. In mild cases the subject is blind and incapacitated for duty for about three weeks. In serious cases for a considerably longer period. The only defense is prompt use of gas mask and even this only guarantees a reduction in losses.

RIOT CONTROL AGENTS

Riot control agents and their less-lethal counterparts are the primary focus of this work. Although references about the military value of lacrimators (such as CN) can be traced to as early as 1887,[13] riot control agents didn't come into widespread civilian law enforcement use until the mid-1960s when police agencies in this country began to routinely use such chemical products as CN and CS against rioting crowds and disorderly individuals. Considered the "most effective, safest, and most humane method of mob control" by the U.S. attorney general in 1968, less-lethal riot control agents were embraced by all levels of law enforcement during those tumultuous years.

Many of these agents have proved extremely effective at producing only temporary casualties. However, prolonged exposure to any of these agents,

especially indoors, can cause serious illness or death.

It must always be remembered that these agents are classified only as less lethal—that does not mean that they are nonlethal.

A few of these agents, such as DM and CNS (see below), are no longer commonly employed because they produce effects considered too severe for use against civilians or in military training.

A brief description of the various riot control agents and their effects follows.

Vomiting Compounds

Adamsite (DM)

Among many of the early agents adapted for employment against civilians during riot or disturbance situations was diphenylaminochloro-arsine, or DM.

DM, sometime referred to as "knockout gas" or "sickening gas," was developed during World War I by Maj. Roger Adams[14] and later used by some U.S. law enforcement agencies. This agent, which may still be found in outdated police department inventories, is generally considered to be unsuitable for police use because of its severe effects. Projection vomiting, instant diarrhea, and intense headaches are induced and may last for days.

NOTE: DM is classified as a sternutator, blood, or nauseating agent, and exposure to it can cause serious injury or death! In fact, the effects are so severe that there have been reported incidents where people exposed to it have committed suicide to relieve their suffering. So, if you are a departmental armorer or chemical agent officer, you would be well advised to inventory any outdated chemical munitions bearing any markings that identify them as DM and dispose of them.

Of course, disposal is another matter, and one that you must make arrangements for based on applicable state, federal, and environmental laws and regulations. Just be sure to take the utmost care with any outdated munitions when handling or disposing of them, particularly if you run across some marked, color-coded green, or otherwise indicated as DM either by name or the chemical formula $NH(C_6H_4)_2AsCl$. There is no known first-aid treatment for exposure to the produced yellow-

Beware of any outdated munitions you may encounter, particularly if you run across any that may be DM.

orange (or sometimes greenish) cloud that smells of licorice or "smoky wood."

Other vomiting compounds include diphenyl-chlorarsine (DA) and diphenylcyanoarsine (DC).

Be advised that none of these vomiting agents should be used for military training or civilian law enforcement use!

Tear-Producing Compounds

CN

Chloroacetophenone, or CN ($C_6H_5COCH_2Cl$), was developed in 1869 by a German chemist named Karl Graebe. Graebe had once been a student of Adolf von Baeyer. Both Baeyer and Graebe conducted research into industrial dye-related materials. This research and the chemicals involved provided them and other chemists of the time with the knowledge and ability to modify and employ these materials as CAWs.

Unlike DM, CN in many forms is still in use today and may be found in modern police arsenals. CN, which gives off a fragrant, apple-blossom-like odor, is a classic lacrimatory agent that acts practically instantaneously, affecting the upper respiratory passages and causing severe irritation of the tear ducts and eyes. These effects have caused

CN and other lacrimator agents to become widely known as tear gas. CN's color code is red.

One important consideration regarding the use of munitions loaded out with CN is the fact that many of them were produced between the 1940s and 1960s.[15] *Many department arsenals still stock munitions from this period.*

In my 14 years of police service, I have used and trained with some CN gas that had been on the job at least 10 years longer than I had. After speaking with numerous police special operations officers from all over the United States, I have found that my experiences are not unique. When you consider that some tactical unit personnel carry all their gear with them all the time (often in the trunks of their vehicles)—subjecting the old canisters to both extreme heat and cold—it is not surprising to learn that the canisters occasionally explode when used.[16] That's one reason why most of the major manufacturers and chemical agent suppliers recommend that chemical munitions be replaced every five years and that older agents be used for training.

Those of us actively employed in the police industry realize that this is not always done. We must therefore make a concerted effort to do what we can to ensure the greatest safety margin for ourselves, our people, and the citizens we serve in regard to our agency's chemical agent stockpiles. If that means notifying administrators—in writing—of the need to refresh chemical agent supplies as well as to provide needed training to department members, then that is what we must do.

Remember, the administrator's consuming fear of the liability monster *can be* effectively used against him—just as a way of leveraging him toward doing the right thing, of course. And, as most of us realize, deep down that is exactly what most administrators truly want to do anyway. By creating a responsibility-illuminating paper trail, we simply light the way for them to follow, allowing them the means to both justify their actions and defend the expenditure of funds for training and equipment. The trick, of course, is to pull it off more than once without getting transferred to Outpost #3.

CNB

CNB is a solution of CN in benzene and carbon

tetrachloride. A powerful lacrimatory agent, it was adopted in 1920 and used until it was replaced by CNS. Unlike those of CNS, the effects of this agent are not long lasting. Like CNS, it is considered obsolete.

CNC

CNC has no specific chemical name. It is simply a solution of CN in chloroform. CNC is considered to be an effective irritating agent. It is not toxic. Its effects, which are short term, include stinging of the skin, irritation of the respiratory system, and uncontrolled flowing of tears.

CNS

Like CNC, CNS has no specific chemical name. It is a mixture of CN, the irritant chloropicrin (PS), and chloroform. Unlike CNC, CNS has been deemed obsolete because it produces extremely severe effects that may persist for weeks after exposure. This makes it highly undesirable for use as a training or riot control agent.

CA

CA, or 4-bromobenzylcyanide, is a lacrimator that smells like fruit and produces a burning sensation of the mucous membranes, severe irritation of the eyes, and intense pain behind the forehead. CA can be persistent and is extremely corrosive on all common metals except lead. Also known as BBC, larmine, and camite, CA is considered obsolete.

CS

CS, or orthochlorobenzalmalononitrile, also known as O-chlorobenzylidene malononitrile or $ClC_6H_4ChC(CN)_2$, was developed by the British[17] in the late 1920s and adopted by both the British and U.S. militaries in the mid-1950s. The first tactical civil use of CS occurred in 1961 on Cyprus.

U.S. police departments began using the agent with the pepper-like odor in the mid-1960s because it was found to be more effective than CN yet less potentially lethal in the event of prolonged exposure.

CS is classified as an irritant, producing extreme burning of the eyes, coughing, runny nose, tightness in the chest and throat, dizziness, and a stinging sensation on moist skin. Heavy concentrations of the agent may also cause nausea and vomiting. Not as fast acting as CN, CS still produces rapid effects and is also more chemically stable in high temperatures, as well as more potent. The color code for CS is blue.

CS1

CS1 is specially formulated to increase the effectiveness and prolong the persistency of CS. CS1 is a micropulverized agent powder consisting of 95-percent crystalline CS blended with 5-percent silica aero gel. This formulation reduces agglomeration and achieves the desired respiratory effects when dispersed as a solid aerosol.

CS2

CS2 is CS that has been blended with silicone-treated silica aero gel. This treatment improves the physical characteristics of CS by reducing agglomeration and hydrolysis, in effect making it water repellant. CS in this form is extremely persistent. When a surface that has been contaminated with CS2 is disturbed, the agent will re-aerosolize and cause respiratory and eye effects even months after the initial application.

CSX

CSX is a form of CS developed for dissemination as a liquid rather than as a powder. One gram of powdered CS is dissolved in 99 grams of trioctylphosphite (TOF). As with CS, CSX stings and irritates the eyes, skin, nose, throat, and lungs of those exposed to it.

CR

CR, also known as dibenz-(b,f)-1,4-oxazepine, is an irritant developed in 1962 by British chemists. It is reportedly more effective than CS, as well as less toxic. In pure form CR is a yellow powder. For riot control use, this powder is dissolved in a solution of propylene glycol and water to form a 0.1-percent CR solution. Eye pain, discomfort, and excessive tearing occur with sometimes painful sensitivity to strong light or temporary blindness. Symptoms generally persist for up to 30 minutes. It was used in Northern Ireland in the 1973–1974 period, where it was called "fire gas" by the media because of the burning sensation it caused to the skin of rioters. Its color code is violet.

PS

PS, also known as chloropicrin, is an irritant used as a riot control agent as well as in commercial industry as a disinfectant and fumigant. PS produces severe sensory irritation in the upper respiratory passages.

Chlorine

Chlorine is a powerful irritant that quickly affects the respiratory tract as well as the skin and eyes. It has a pungent odor. Its effects are generally short term, though exposure to it can cause severe damage to the mucous membranes of the respiratory passages. Excessive inhalation may cause death. Chlorine was first used as a chemical weapon in World War I.

Although not used today as a chemical weapon by either the military or the police in the United States, chlorine is included here because of its widespread availability and common commercial and domestic use. It is used to treat drinking water and is found in household bleach. An officer I spoke with was once sprayed in the face at close range with a mixture of chlorine bleach and water while responding to a domestic disturbance. The female subject who sprayed both him and his partner used a small pump-type spray bottle. According to the officer, it was effective, causing severe irritation of the eyes and respiratory tract. The suspect was arrested and charged with assault and battery with a dangerous weapon.

OC

Oleoresin capsicum is classified as an inflammatory agent. It usually causes an immediate swelling of the mucous membranes, closing of the eyes, and a sensation of intense burning of the skin and mucous membranes inside the nose and mouth. When delivered in atomized form it may also be inhaled. This causes additional effects such as uncontrollable coughing, gagging, and gasping for breath.

The three major ingredients in OC that determine its strength—capsaicin, dihydrocapsaicin, and nordihydrocapsaicin—are called capsaicinoids. These capsaicinoids are derived directly from cayenne chili peppers. Although first synthesized successfully in the 1930s, OC remained an obscure agent for many years because there were no suitable

carrier/vehicle and propellant to deliver it. That changed in 1974 with the introduction of Cap-Stun.

The law enforcement industry, long-time users of chemical agents such as CN and CS, were slow to adopt this new product. Then in 1989, the Firearms Training Unit of the FBI Academy in Quantico, Virginia, completed an intensive three-year OC testing and research program. It was found that OC was effective on most subjects (current statistics indicate an approximate 72–84 percent effectiveness margin) and that it produced no harmful aftereffects or permanent damage. The FBI then adopted the product, authorizing its use by its agents and special-response teams.

As often happens, what the FBI does, so does the bulk of U.S. law enforcement. OC was off and running. As with many new products, OC was also somewhat overrated in the beginning, with one advertisement going so far as to depict a uniformed officer lying on the ground, baton beside him, weapon holstered, spraying his OC at a lunging attacker armed with a knife.

We in the law enforcement community now know through painful experience that attempting to control by less-lethal means a subject presenting a lethal threat is neither realistic nor prudent under most circumstances.[18]

HC

Hexachlorethane titanium (HC) and tetrachloride anthroquinone are used to produce obscuring smoke. The smoke results from the combustion of a mixture of HC, grained aluminum, and zinc oxide. Produced in different colors, smoke is generally used as a signaling tool or a means of providing concealment for movement of personnel, though it has little effect on modern thermal sights. Combined with a CAW grenade, it can also be used to "float" the chemical agent, increasing both the effectiveness and area of dissemination.

Although not technically classified as a CAW, HC should not be used indoors because it will displace oxygen. And since standard respirators do not provide oxygen but simply purify contaminated oxygen, suffocation is possible even when the person exposed to HC is masked. Having been exposed to quite a bit of the mildly irritating smoke in military and police settings, I was somewhat surprised when I

Smokes.

Expired munitions.

discovered that petroleum-based HC has been determined to be carcinogenic in heavy concentrations. Of course, just about *anything* is carcinogenic in heavy enough concentrations. All the same, if you're involved in any type of operations that entail the use of "just smoke," as we used to call it, you're well advised to avoid breathing the vapors. HC's color code is yellow.

Saf-Smoke

Federal Laboratories produces a smoke product made with purified terephthalic acid (1,4-benzenedicarboxylic acid). Saf-Smoke produces a vivid white smoke that the company describes as *essentially* (author's italics) nontoxic.[19] According to the manufacturer, Saf-Smoke has been declared not hazardous/noncarcinogenic by OSHA. It too will displace oxygen, however, and should not be used indoors. Its color code is also yellow.

OUT-OF-DATE MUNITIONS

As noted above, it is not uncommon to come across expired or obsolete chemical agent munitions. If these munitions or agents are used (as they frequently are in the real world) and nothing unusual or unexpected occurs during their use, then chances are nothing bad will come of it.

If, however, something (read *anything*) does go wrong and someone is injured unnecessarily or unintentionally, be advised that the fact that you or your department used expired or obsolete munitions

may result in a very large checkmark in the "bad thing" column. This is true regardless of whether the injury occurred during training or while officers were engaged in an actual mission.

Even if the outdated chemical agent munitions played no part in the actual injury, just the fact that you were using obsolete or expired munitions could be used to indicate to the jury an overall lack of professionalism or competence on your part. This would obviously exacerbate any appearance of culpability as well as liability on the part of you or your agency in the eyes of the judge, jurors, onlookers, and media. It just isn't worth it.

So again, if you are maintaining any expired munitions, you would be well advised to properly dispose of them and restock your inventory. (One way to deal with the problem of maintaining a current inventory without having to continually spend large amounts of hard-to-come-by funds is presented for consideration in Chapter 19.)

For your information, the following companies either no longer exist or produce chemical agent munitions. If you possess any chemical munitions bearing their logos or names, you can pretty much be assured that they fall under the heading of "out-of-date munitions."

- Lake Erie
- Penguin Associates, International
- Smith & Wesson Chemical Company
- LECCO
- General Ordinance Equipment Corporation

Note the corrosion visible on this CS grenade. Grenades that are extremely old or improperly stored present hazards both to the departments that possess them and the individuals who may employ them. Gas grenades in similar condition have reportedly exploded when employed.

NOTES

1. Chemical weapons are classified as toxins and lethal and incapacitating agents.
2. Chemical agent weapons refer to those classified as riot control agents.
3. "The Contracting Powers agree to abstain from the use of projectiles the object of which is the diffusion of asphyxiating or deleterious gases." Germany was a signatory to this article of the Hague Convention.
4. Lt. Col. George Christopher et al. *U.S. Army Bio Warfare Handbook—Types, Risks, Precautions.* (Ft. Detrick, Md.: U.S. Army Medical Research Institute of Infectious Diseases, 1998).
5. In October 1914, January 1915, and March 1915, the Germans launched chemical attacks against Allied forces with no noticeable effect. Improper preparation of the munitions and extremely cold weather are believed to be the causes of the failure.
6. Fritz Haber, considered the "father of modern chemical warfare," served as a consultant to the German War Ministry and organized gas attacks and defenses against such attacks. Haber was awarded the Nobel Prize in chemistry in 1918 for his work on fixing nitrogen from the air.
7. Gen. John J. "Black Jack" Pershing recorded an eyewitness account of this attack in his memoirs, *My Experiences in the World War* (New York: Frederick A. Stokes Company, 1931). Account witnessed by Sir John French and reported in his dispatches to Lord Kitchener (p. 166, vol. 1).
8. O-ethyl S-2 (diisopropylaminoethyl) methylphosphonothioate.
9. In 1968, the U.S. Army reportedly tested the effects of VX by spraying 20 pounds from an E-4 Phantom jet at Dugway Proving Ground, Utah. Over 6,000 sheep were killed within a 25-mile radius of the drop.
10. It is believed that a sodium cyanide agent may have been used by terrorists in the 26 February 1993 New York Trade Center bombing, but the agent burned during the explosion instead of vaporizing, sparing untold thousands of lives.
11. Devised by the American chemist Dr. W. Lee Lewis in 1917. Unlike mustard, lewisite produced severe pain upon first contact with the skin.
12. H is European usage.
13. The renowned German chemist, Professor Adolf von Baeyer, reportedly referred to the military value of lacrimators in his lectures to advanced students as early as 1887.
14. The agent was named adamsite in his honor.
15. A mixture of CN, various petroleum-based carriers, and a mixed freon-hydrocarbon solvent was used to produce chemical Mace in the late 1960s.
16. Though rare, this does occasionally occur, especially with CN-loaded pyrotechnic grenades (CN melts at a lower temperature than CS: 129.2 to 131°F for CN, in contrast to 199.4 to 205.8°F for CS). At these temperatures, the agents may melt and block emission ports, causing excessive pressure buildup when the grenade is ignited. In severe cases the blockage will result in the canister's fragmenting.
17. British chemists B.B. Carson and R.W. Stroughton. The name "CS" is reportedly derived from the scientist's last names.
18. On 8 April 1992, Frank Ward, a police officer with the John Day Police Department in Oregon, employed OC against an attacker who was armed with a heavy log. Officer Ward emptied nearly the entire can of chemical spray, directly hitting the facial area of his attacker, Sidney Dean Porter. Porter, however, continued the assault and beat Ward to death with the log.
19. Some minor adverse effects have reportedly been observed during testing on laboratory rats.

Quick Listing Sheets for Today's Chemical Agent Weapons

NOTE: In keeping with the overall intent and purpose of this book, only the use and employment of currently accepted, less-lethal classified CAWs will be discussed from this point forward. As of this writing, only CN, CS, and OC are considered appropriate CAWs for law enforcement use in the United States.

This chapter is intended to provide an easily accessed, less-lethal CAW reference guide. To this end, a quick listing sheet (QLS) for CN, CS, and OC, as well as a usage key to explain the terms used in the sheets, is included on the following pages.

CN, CS, and OC are generally available in solid, micropulverized,[1] or liquid form. These agents are all currently employed throughout the world, primarily by police, military, and security personnel to control or redirect disorderly persons or rioting groups during civil disturbances. The agents are also used for training purposes to prepare personnel to operate in chemically affected areas. (An overview of these agents is provided in Chapter 2.)

CN (QLS-1)

1. Abbreviation of chemical name: CN.
2. Chemical name: Chloroacetophenone.
3. Color code: Red.
4. Classification: Lacrimator.
5. Physical state: Microparticulate solid.
6. Odor: Apple blossoms.
7. Median lethal dose: 14,000 (mg-min/m^3).
8. Median incapacitating doge: 20 (mg-min/m^3).
9. Reaction time: 1–3 seconds.
10. Duration of effect: 30–45 minutes.
11. Physiological effects: Temporary severe eye irritation causing profuse tearing and voluntary closing of the eyes; mild nasal, respiratory tract, and skin irritation.
12. Psychological effects: Mild panic.
13. Failure rate: Approximately 40 percent on humans, especially emotionally disturbed persons or subjects under the influence of drugs or alcohol.
14. Protection required: Protective mask.
15. Stability: Stable.
16. Decontamination:
    a. Leave contaminated area.
    b. Face into the wind.
    c. Remain calm.
    d. Blow nose.
    e. Keep eyes open as much as possible and do not rub. Let eyes tear freely because this will carry agent out. If agent particles are rubbed into eyes, flush with large amounts of clear water. Generally, all major effects will subside in 10–20 minutes. NOTE: If effects worsen or do not subside within specified time, seek medical attention!
17. Use: Training and riot control agent.

CS (QLS-2)

1. Abbreviation of chemical name: CS.
2. Chemical name: Orthochlorobenzylmalononitrile.
3. Color code: Blue.
4. Classification: Irritant.
5. Physical state: Microparticulate solid.
6. Odor: Peppery.
7. Median lethal dose: 25,000 (mg–min/m^3).
8. Median incapacitating dose:[2] 10–20 (mg–min/m^3).
9. Reaction time: 3–7 seconds.
10. Duration of effect: 30–45 minutes.
11. Physiological effects: Temporary severe eye irritation causing profuse tearing and voluntary closing of the eyes; severe nasal, respiratory tract, and skin irritation. Tightness in chest/airways/throat. Dizziness and swimming of head. Heavy doses may cause nausea and vomiting.
12. Psychological effects: Mental disorientation, confusion, panic.
13. Failure rate: Approximately 40-percent failure rate on humans, especially emotionally disturbed persons or subjects under the influence of drugs or alcohol.
14. Protection required: Protective mask.
15. Stability: Stable.
16. Decontamination:
    a. Leave contaminated area.
    b. Face into the wind.
    c. Remain calm.
    d. Blow nose.
    e. Keep eyes open as much as possible and do not rub! Let eyes tear freely to carry the agent out. If agent particles are rubbed into eyes, flush with large amounts of clear water. Generally, all major effects will subside in 10–20 minutes. A solution of 5–10 percent sodium bicarbonate may be used on skin in small amounts if desired. NOTE: If effects worsen or do not subside within specified time, seek medical attention!
17. Use: Training and riot control agent.

OC (QLS-3)

1. Abbreviation of chemical name: OC.
2. Chemical name: Oleoresin capsicum.
3. Color code: None or orange.
4. Classification: Inflammatory.
5. Physical state: Micropulverized solid/oily resin (most cases).
6. Odor: None.
7. Median lethal dose: Undetermined[3] (mg–min/m^3).
8. Median incapacitating dose: Undetermined (mg-min/m^3).
9. Reaction time: 1–2 seconds.
10. Duration of effect: 30–45 minutes.
11. Physiological effects: Temporary severe eye inflammation causing involuntary closing of the eyes. Severe burning and inflammation of mucous membranes. Severe burning and inflammation of throat and respiratory tract if inhaled, allowing only survival breathing. Severe burning and inflammation of skin.
12. Psychological effects: Intense panic, loss of desire and will to fight, confusion.
13. Failure rate: Approximately 16–28 percent failure rate on humans.
14. Protection required: Protective mask.
15. Stability: Stable.
16. Decontamination:
    a. Remove contaminated clothing.
    b. Flush with copious amounts of water.
    c. Remain calm.
    d. Blow nose.
    e. Keep eyes open as much as possible and do not rub! Let eyes tear freely to carry agent out.
    f. Do not apply any creams, balms, salves, or oils to skin! Generally, all major effects will subside in 20–45 minutes. A solution of COOL-IT! or similar product may be used on skin in small amounts if desired.

    NOTE: If effects worsen or do not subside within specified time, seek medical attention!

17. Use: Individual and riot control agent.

QLS USAGE KEY

1. Abbreviation of chemical name: Used as designation.
2. Chemical name: It is common to find slightly different spellings for the agents, particularly for CS and CN.
3. Color code: Used on canisters and munitions to indicate agent contained within. May appear as colored bands or stickers on individual containers.
4. Classification: Indicates categorization of agent. This is based on the agent's properties and effects.
5. Physical state: Indicates form of prepared agent. For example, CN and CS are white crystal-like solids. When prepared for pyrotechnic[4] or aerosol use, each crystal is approximately 4–8 microns in size.[5] In this form the agent is described as a microparticulate solid. When prepared for use in a blast or expulsion munition, the crystals, again 4–8 microns in size, are combined with other inert materials[6] that prevent the agent from "caking" and aid in dispersal. In this form the agent is described as a micropulverized solid. OC is different from CN and CS in that it is not in crystal form. Rather, it is an oily resin that is dissolved in a liquid carrier.
6. Odor: Smell given off by burning or atomized agent.
7. Median lethal dosage (LCT_{50}): The LCT_{50} of a chemical agent is the concentration multiplied by the time of exposure that is lethal to 50 percent of exposed personnel. The unit used to express LCT_{50} is milligram minutes per cubic meter (mg-min/m^3). For example, CN (reference QLS-1) has an estimated LCT_{50} value of about 14,000 mg-min/m^3.
8. Median incapacitating dosage (ICT_{50}): The ICT_{50} of a chemical agent is the concentration multiplied by the time of exposure that will incapacitate 50 percent of exposed personnel. Again, the unit used to express ICT_{50} is milligram minutes per cubic meter. Using CN (reference QLS-1) again, we can see that the estimated ICT_{50} value is about 20 mg-min/m^3. (A more detailed explanation of both LCTs and ICTs is included in Appendix A.)
9. Reaction time: The estimated amount of time required for the agent to take effect once a subject has been properly exposed to it.
10. Duration of effect: The estimated amount of time that the effects will continue once a subject has been properly exposed to an agent.
11. Physiological effects: Types of physical effects or reactions that may be experienced by a living organism after exposure to a particular chemical agent.
12. Psychological effects: Types of psychological or mental effects that may be experienced by a living organism after exposure to a particular chemical agent.
13. Failure rate: Approximate rate of failure of an agent to induce sufficient physiological and psychological effects to result in incapacitation of a subject properly exposed to that agent.
14. Protection required: Amount and type of protection required to protect a subject from the effects of an agent. The amount and type of protection required is based on field concentration levels of exposure.
15. Stability: Stability refers to the chemical stability of the particular agent while in storage. (Proper storage conditions are described in Chapter 19.)
16. Decontamination: Describes recommended methods that should be used to assist in decontaminating subjects that have been exposed to the various agents. (Decontamination of clothing, property, and buildings is addressed in Chapters 11 and 17.)
17. Use: Recommended use and applications of individual agents.

NOTES

1. Reduced to an extremely fine powder or dust (1 micron = 1/25,000 of an inch).
2. It is recommended that the ICt_{50} value of 10 always be used for CS applications.
3. No median lethal or median incapacitating doses have been determined for OC as of this writing. It is believed, however, that it would take an enormous amount of OC agent to produce fatal results.
4. As a rule, agent dissemination by burning produces a much smaller particle than that produced by blast dispersal.
5. Refers to the particle size of the crystallized agents that are carried in the chemical cloud or liquid carrier agent. A micron is 1/25,000 of an inch in size. Particles smaller than 1 micron in diameter are referred to as submicron particles..
6. Referred to as buffering agents.

Methods of Delivery:
An Overview

CAWs can be delivered to the desired objective area by several means. This chapter briefly outlines the currently available methods of delivery, with following chapters going into detail regarding the specific uses and applications of each.

Before beginning, however, I feel it is worth explaining my decision to employ the term *objective area* instead of *target*.

A WORD ABOUT WORDS

Many would argue that the use of the term *objective area* instead of *target* is motivated by current trends of political correctness, but I can assure you this is not the case—not entirely, that is. The reality is that the members of U.S. law enforcement agencies are currently working in an environment, on the streets as well as in the courtrooms, that is often hostile to us.

As has been shown in recent years, we in law enforcement may well win a battle on the streets only to lose the war in the courtroom. Even though we realize that the physical battle is the more important, the legal battle can still prove

extremely costly to us in terms of mental anguish and monetary loss. Those of us operating in the post–Rodney King era must also consider the fate of the officers involved in that action and realize that if the social climate is sufficiently heated, our lawful actions may even result in the loss of our personal freedom (read *prison*).

That is why the simple choice of terms we use to describe our reasoning and actions in written reports and verbal testimony—insignificant as it might first appear—is actually critical to our total survival.

The selection of technically correct, unincendiary verbiage is also easy to incorporate and costs us and our departments nothing other than the changing of a few words in written policies, procedures, and training literature. Failure to take such simple action, however, can be devastating, especially if you ever have to face a zealous attorney who would enjoy nothing better than spitting out sentences such as, "So you *targeted* my client, didn't you? You targeted him and went after him and filled his house with GAS! Isn't that what it says right here in your report? The TARGET?"

Now even though I know, you know, and, yes, even the attorney knows that the twisting of words and the histrionics are not based in fact or reality, the truth is that the *jury* might not know.

My point here is for you to avoid providing the manipulating, lowdown, snake-in-the-grass—sorry, I mean "legal professional"—with any ammunition, especially in the form of your own words taken from your own report.

This applies to any word or phrase that could, because of its connotations, be used to make us look overaggressive or unprofessional. "Shooting to stop" as opposed to "shooting to kill" is a prime example of how we can alter our wording to more clearly and accurately describe both our actions and intent.

Simply by choosing this type of language, we will be better prepared to engage an attorney/adversary in a war of words and perceptions on his terrain.

The idea of preparing for an engagement long before it occurs is not new. Twenty-five centuries ago, Sun Tzu wrote in *The Art of War*:

The general who wins a battle makes many calculations in his temple before the

battle is fought. The general who loses a battle makes but few calculations beforehand. Thus do many calculations lead to victory, and few calculations to defeat; how much more no calculation at all! It is by attention to this point that I can foresee who is likely to win or lose.

Bottom line: Study the situation; know the terrain; plan ahead. Give your adversaries nothing to use against you!

GRENADES

While on the subject of giving an adversary nothing to use against you, one word comes immediately to mind: *grenade*. Any type of grenade.

I remember being very impressed the first time I handled an M33 fragmentation grenade in the U.S. Army. What impressed me even more was the way my drill instructor stood calmly watching for a few seconds, head and shoulders exposed, after I had pulled the pin and tossed the little green ball over the protective wall that stood between us and the impact area. As he slowly crouched down next to me just as the grenade detonated, he said, "Good hit," and then drawled, "Conni, you ever see them war movies where a grenade rolls into the foxhole and the guy jumps on it?"

"Yes, Sergeant," I nodded, glad he wasn't yelling at me, and waited for him to extol the virtues of self-sacrifice for my benefit.

"Well, that guy's an asshole. If you've got time to jump on it, you've got time to throw it back."

This philosophy definitely has merit. Of course,

the success of either technique will depend on the type of grenade and fuse used, as well as the reflexes and training of the recipient.

In regard to CAWs, it is commonly known that during the riots of the 1960s and 1970s, various types of chemical agent grenades employed by the police were picked up, marked "RETURN TO SENDER," and tossed back by protesters outfitted with gas masks and oven mitts.

The riots of the 1980s, 1990s, and the new millennium proved that the need for effective, high-volume-producing, less-lethal CAWs still exists, as does the need for more efficient (and less easily returned) delivery systems.

Current manufacturers of CAWs have been making progress in this area, incorporating various design improvements and modifications into the munitions. Some of these designs, such as the Triple Chaser (see Chapter 5), have been around for a while, while others, such as AERKO International's CLEAR OUT irritant grenades (see Chapter 7), have come on the scene fairly recently.

Pyrotechnic

The pyrotechnic grenade, in service for nearly 80 years, is still in pretty much the same form as it first appeared. This type of combustion grenade generally puts out a lot of heat as well as flames and agent, and if used indoors may make one glad that

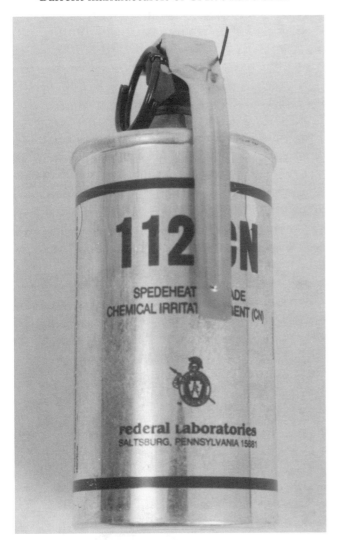

Federal Laboratories' 112CN Spedeheat pyrotechnic grenade.

Federal Laboratories' 121CN "instantaneous-blast-dispersion" grenade. After the pin is removed and spoon released, a delay element is ignited. Once burned through, a small explosive charge is ignited, causing the grenade body to rupture along scored "emission grooves" and blasting the micropulverized agent into the atmosphere.

the fire departments in most parts of our country are so quick to respond. These grenades, requiring from 10 to 150 seconds to disperse their payloads, are the type that influenced peace-loving—though decidedly militant—anarchists to accessorize with their mom's oven mitts.

Expulsion

The expulsion, instantaneous blast, or blast dispersion grenade uses no fuel, flames, or heat. Rather, micropulverized agent particles are mixed with a buffering agent that keeps the load in powdered form. Upon detonation, the agent is released virtually instantaneously into the atmosphere by one of the following methods:

- A propellant-driven piston is used to expel finely ground particles of the agent directly into the atmosphere.
- The instantaneous release of compressed gas propels the agent into the atmosphere.
- A small explosive charge surrounded by micropulverized agent is detonated, expelling the agent into the atmosphere through ports in the grenade body. Some grenades of this type also employ perforated or scored cardboard, plastic, or aluminum bodies that burst when the charge is detonated, releasing the agent as well as possibly producing fragmentation.

These types of devices are generally considered appropriate for indoor use because they produce no flames and/or extreme heat.

Aerosol

Aerosol grenades have also come into their own in the past decade, especially since the advent of OC. These grenades tend to disperse their entire payload in a matter of seconds, negating the possibility of their being thrown back. Most of these devices use nonflammable solvents in pressurized containers to deliver the agent in a quickly spreading cloud. Aerosol grenades can be configured like standard CAW grenades with M201A1-type fuzes or industry-standard aerosol cans, as used by Aerko International.

PROJECTILES

Pyrotechnic

Pyrotechnic projectiles are usually launched from a gas launcher designed for this purpose. Pyrotechnic projectiles generally range in size from 6 to 12 inches and can cause severe injury or death if the projectile itself strikes a human being. It is critical to remember, especially when firing the long-gun-like gas launchers, that the idea is to deliver the

CLEAR OUT aerosol grenade.

agent to the objective area by way of the projectile—not, as with the firearm, to deliver the projectile into the body of a person.

Of course, if the situation ever arises where it is necessary to stop an immediate threat to save your life or someone else's, then you would be required to justify your use of deadly force, as in any other such case.

There is also a very definite fire hazard associated with the use of these munitions.

A 12-gauge nonpyrotechnic projectile will penetrate hollow-core doors and windows. The Defense Technology No. 23 rounds shown here are highly accurate.

Nonpyrotechnic

Nonpyrotechnic projectiles are also available in several forms. These munitions are generally configured as liquid-load projectiles; dry-powder-load projectiles; and muzzle-blast dispersion rounds. There is no fire hazard associated with the use of any of these types of munitions. All are efficient and effective when employed properly.

Having been involved in hostage/barricade situations where 12-gauge barricade-penetrating projectiles similar to those shown in the photo (left) were used, I can say I have great confidence in them. Barricade-penetrating projectiles are fired through either a 12-gauge shotgun or a 37, 38, or 40mm gas launcher.

Upon striking a solid object, the projectile disintegrates, dispersing a concentration of CS, CN, or OC in a fine aerosol or in powdered form.

As with the pyrotechnic projectiles, nonpyrotechnic projectiles can cause severe injury or death should they strike a human being.

FOG GENERATORS

Fog generators are the big boys on the chemical agent block. Imposing in appearance, extremely cost-effective and efficient, they are a good investment if your department is (or may be tasked with) dealing with large crowds of unhappy people.

Curtis Dyna-Fog Golden Eagle thermal fog generator.

An ISPRA compressed-gas projector.

When properly used, the foggers generally tend to make these people even more unhappy but will also cause them to go be unhappy somewhere else: these devices are capable of producing thousands of cubic feet of obscuring smoke or chemical agent gas in minutes.

The foggers may be carried by an individual officer and employed on foot or from inside a vehicle. They can also be mounted directly on a vehicle, armored or otherwise, and activated from within. Foggers are also useful for filling a building with smoke or chemical agents; both the amount and strength of agent that is released can be controlled by the operator.

COMPRESSED-GAS PROJECTOR SYSTEMS

Compressed-gas projector systems provide another nonpyrotechnic method of delivering chemical agents. The systems most commonly used appear and function like common fire extinguishers. The unit is loaded with a formulation of agent in

These two aerosol spray units are produced by Fox Labs, Inc. The unit on the left is equipped with a heavy-stream nozzle. The unit on the right has a fog nozzle.

Small "pocket-size" (center) or "pen-type" (right) ASR units such as those shown here may be useful for some special-purpose applications, buy they are generally not practical for serious law enforcement or tactical field use. Combination devices such as the ASP Key Defender (shown far left with test cartridge) that are designed to appeal to people as multipurpose self-defense tools (e.g., key ring, flail, kubaton, ASR) very often fail to perform effectively while also drawing undue attention to themselves and the people carrying them. So, unless you've got a specific reason for carrying any of these types of mini-ASR devices, you're probably better off choosing and carrying a full-size unit like those shown on the previous page.

powder or liquid form. The dispenser body is then pressurized with nitrogen gas. When the safety pin is removed, a single press of the actuator handle allows the pressurized agent to be emitted down range in

the form of a high-volume cloud or stream. Units of this type are currently seeing extensive use in such places as Israel and Hong Kong.

AEROSOL SPRAYS

Aerosol subject restraint (ASR) sprays are used extensively by the police industry in the United States. The development of an effective liquid carrier for the very effective oily, resin-based oleoresin capsaicum agent has had a great deal to do with this current popularity.

Although chemical Mace aerosol sprays have been used for law enforcement purposes since the mid-1960s, they use formulations containing CN, a product manufactured chemically.[1]

OC, on the other hand, is considered a "natural" product in that its active ingredient is derived from peppers. This fact also makes the OC agent more attractive to administrators and allows it to be placed lower on the use-of-force continuum than other laboratory-manufactured agents. In fact, OC aerosols are often put on the same use level as empty-hand techniques, providing a good option for the line officer if the situation meets the recommended criteria. (See Chapter 11 for more on this.) OC sprays are not, however, to be used as a replacement for higher levels of force, should the situation escalate or require a stronger response.

NOTE

1. Chemical Mace aerosol sprays use formulations containing CN or CN/OC.

The Pyrotechnic Grenade

Above Left: Some of the many pyrotechnic grenades available. With the exception of the "flameless" type of pyrotechnic grenade, they are designed specifically for outdoor use. They must be employed by qualified personnel who have been trained in their proper use. Outdoor crowd control and civil disturbance situations are the primary intended applications.

Above Right: Federal Laboratories 517CS flameless pyrotechnic grenade uses a unique double-walled construction that minimizes the possibility of fire by shielding the surrounding area from the flames. It may, however, generate enough heat to ignite ambient gasses.

STRENGTHS AND WEAKNESSES

Pyrotechnic grenades, first designed more than 80 years ago, are still viable tools for modern law enforcement use. They may be thrown by hand or launched into an objective area. This launching ability—especially if dealing with hostile crowds—is extremely beneficial: officers may stay at a safer distance than if hand-throwing the munitions. This should aid officers in keeping out of range of most rioter-thrown projectiles.[1]

These munitions, when properly distributed, also provide excellent coverage for large outdoor areas. Smoke and CS or CN agent grenades may be used exclusive of one another, depending upon the tactical need and desired effect. They may also be used in tandem to great effect; the smoke issuing from a non-agent-loaded munition will "pick up" and carry chemical agent crystals, multiplying the area coverage. Smoke may also be used alone as a signal or diversion or to conceal the actions or movements of personnel.

One of the major drawbacks of the pyrotechnic design is the fire hazard they present. For this reason, pyrotechnic grenades should never be thrown onto rooftops or into or under buildings or vehicles. Care must also be taken when using these devices outdoors. They should not be introduced into environments containing flammable material.

Because the devices also produce great quantities of oxygen-displacing smoke while simultaneously consuming oxygen during the combustion process, there is also the consideration of oxygen deprivation in enclosed areas.[2]

HOW IT WORKS

The pyrotechnic grenade works by process of combustion. Simply put, the chemical agent or smoke ammunition (in crystallized form) is mixed

with a fuel substance—generally common sugar—and pressed into a cake, pellet, or block configuration.

This material is then loaded into a canister. Some type of fuze device is then activated, and the agent/fuel material is ignited. As it burns, smoke is produced and released through a series of gas (or emission) ports located in the body of the grenade. When using either CS or CN munitions, the chemical agent crystals are carried into the air with the smoke.

Pyrotechnic devices such as these are commonly referred to as continuous-discharge grenades because it can take from 10 to 150 seconds to discharge a payload.

VARIOUS PYROTECHNIC GRENADE CONFIGURATIONS

Shown on the following pages are several variations of commercially available pyrotechnic

A pyrotechnic smoke grenade dispersing its payload. Smoke grenades are generally configured with four emission ports on the top of the canister, while CS and CN grenades have four on top and one on the bottom. Note the flames and sparks produced by the intense combustion. These types of munitions should not be used indoors, thrown on rooftops, or thrown into crawl spaces! Besides the fire hazard they present, these types of munitions also deplete oxygen. They should also be kept away from flammable materials when used outdoors.

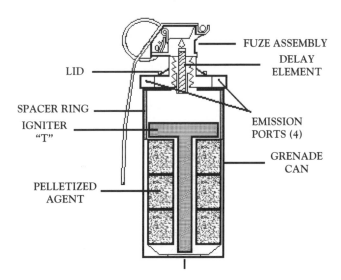

Illustration 5-1. Cutaway diagram of a continuous-discharge-type grenade.

LID

SPACER RING

IGNITER "T"

PELLETIZED AGENT

FUZE ASSEMBLY

DELAY ELEMENT

EMISSION PORTS (4)

GRENADE CAN

grenades. Each is designed to perform in a slightly different way. The primary purpose of all of these devices, however, is simply to deliver the appropriate agent to the strategic area where it will produce the desired effect. Just what these desired effects should be is discussed in Chapters 16 and 17.

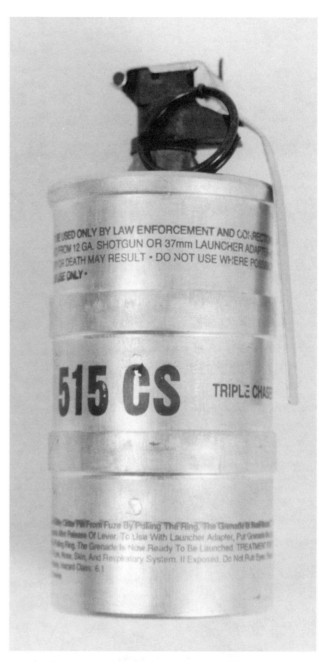

Standard Pyrotechnic Grenade. This model has four gas emission ports on top and one in the base. (As noted in the text, smoke grenades are normally configured with only the four emission ports on the top; CS/CN grenades will also have an emission port in the base.) The grenade may be loaded with CS, CN, or smoke. It can be launched with either a specially adapted gas launcher or a shotgun. Discharge times for CS/CN range from 30 to 40 seconds. Smoke will discharge for up to 2.5 minutes or longer.

Triple Chaser, or Triple Charger, Grenade. This model has four emission ports on top of each canister. It can also be loaded with CS, CN, or smoke. This type of munition consists of three separate aluminum canisters that are pressed together with expelling (or separating) charges between each section. When deployed, the canisters separate, each dispersing its payload as much as 20 feet apart from the others. Very often when thrown by hand, at least one of these canisters will tend to obey Murphy's Law by automatically skittering right back to the hapless officer who threw it. These devices are also launchable. Discharge time for CN/CS or smoke is approximately 20 to 35 seconds.

Rubber Ball Grenade. This type of grenade has four emission ports located around the equator of its body. CN, CS, and smoke loads are available. It has a fast discharge time and also tends to skitter across hard surfaces, making it both entertaining to watch and more difficult to pick up and throw back. The manufacturers of these types of grenades recommend they be kept in their foil pouches until they are to be used because CN or CS may leach through the rubber body. Some models are launchable. These types have a discharge time of approximately 15 to 20 seconds.

Tactical or "Pocket" Continuous Discharge Grenade. This is the smallest of the pyrotechnic grenades currently available. One gas emission port is located on the bottom. It is classified as "tactical" because of its compact design and easy concealability. Discharge time is approximately 20 to 30 seconds.

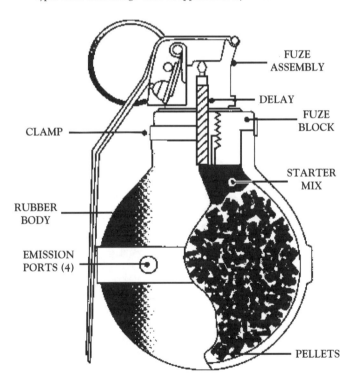

FUZE ASSEMBLY

DELAY

FUZE BLOCK

CLAMP

STARTER MIX

RUBBER BODY

EMISSION PORTS (4)

PELLETS

Illustration 5-2. Cutaway diagram of a rubber ball grenade. (Courtesy of Defense Technology.)

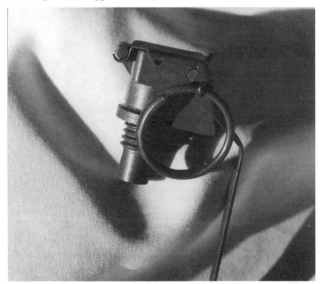

The Fuze. Close-up of M201A1 mechanical fuze.

THE FUZE

The fuze is the critical component of a pyrotechnic grenade. If it doesn't work, the grenade is as useful as a paperweight. Most pyrotechnic grenades use some type of mechanical fuze to ignite the agent.[3]

M201A1 Type Fuze

All of the pyrotechnic grenades illustrated above are equipped with M201A1 type fuzes (as seen in Illustration 5-3). The M201A1 fuze is one of the most common, using a spring-loaded striker. When the safety pin is removed, the striker is driven forward by the action of the spring. The safety lever, unless it is held down, is then cast off by the striker, exposing the fuze's percussion primer located beneath it. The striker's faceplate, shaped like a firing pin, then hits the primer, igniting a delay element (as shown in Illustration 5-4). Once the delay element is burned through, (generally 1.5 to 5.0 seconds, depending on the configuration), the ignition mixture detonates, and the payload's fuel begins to burn.

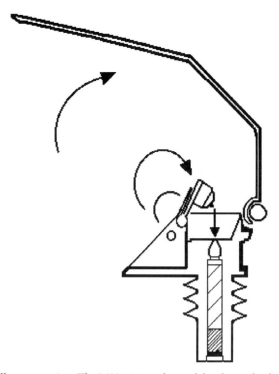

Illustration 5-4. The M201A1 mechanical fuze being fired.

Percussion Mechanical Fuze

Another type of effective fuze is the percussion mechanical fuze. If you are familiar with the discontinued Smith & Wesson "Mighty Midget," you are familiar with this device. The most notable difference between this fuze and the M201A1 is that it uses a manual striker instead of a spring-driven

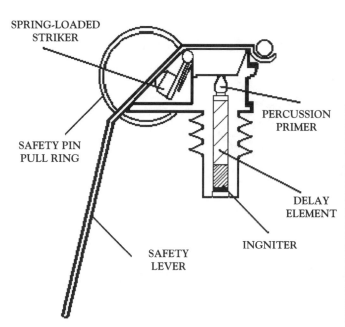

SPRING-LOADED STRIKER

PERCUSSION PRIMER

SAFETY PIN PULL RING

DELAY ELEMENT

INGNITER

SAFETY LEVER

Illustration 5-3. M201A1 mechanical fuze.

Percussion Mechanical Fuze. Close-up of percussion fuze.

striker. Once the safety pin has been pulled, the manual striker is driven against a hard object. This causes the striker's firing pin to hit the primer, which ignites the delay element. After approximately 3 seconds, the ignition mixture is detonated and the agent is dispersed.

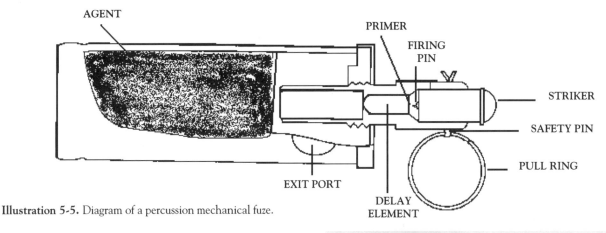

Illustration 5-5. Diagram of a percussion mechanical fuze.

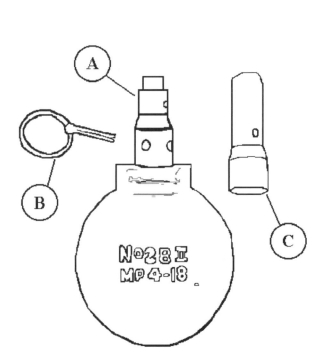

Illustration 5-6. The percussion mechanical fuze as it was used during World War I with the 1917 No. 28 Mk II tear gas grenade. This gas grenade could be thrown by hand or catapult. Both the percussion fuze (A) and protective cover (C) were made of brass. Once the safety pin (B) was pulled, the operator would drive the striker against a hard object, igniting the delay element, and the grenade would be delivered into the objective area. Grenades such as these were employed in clearing confined trenches and dugouts.

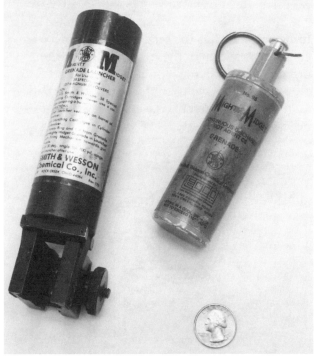

Mighty Midget. The Smith & Wesson "Mighty Midget" grenade (right) and launching cup. The politically incorrect name should be the first indicator that this particular product hasn't been produced in the United States for some time. Driving the striker against a solid object ignited the delay element. This ignition system also allowed for the device to be launched from a revolver with the addition of a special launching adapter cup. The grenade was inserted into adapter cup with the fuze facing the muzzle of the gun. When the blank firing round loaded into the pistol was discharged, the striker was driven forward simultaneously as the grenade was launched. (Grenade and launcher from author's collection.)

EMPLOYMENT OF THE PYROTECHNIC GRENADE

How to Hold It

The grenade should be held firmly in one (the dominant) hand, with the safety lever (or spoon) pressed into the web of the hand as shown. The primary arm should be extended to the front of the body. Grenades should never be lifted or handled by the pull ring!

Pyrotechnic rubber-ball-type grenades should be held in similar fashion, with the safety lever (or spoon) pressed into the web of the hand. Left-handed individuals may opt to hold grenades inverted, spoon in the web of hand, pull-ring down, when deploying them.

Pulling the Pin

Tom Robbins demonstrates one simple and effective method for pulling the pin cleanly and quickly from the fuze assembly. Note that he does not use his teeth. Keeping the grenade extended to the front, grasp the pull-ring securely with the index finger of the nondominant hand, palm forward, thumb-side down as shown and . . .

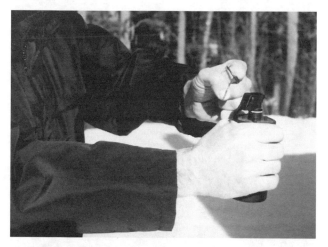

. . . simultaneously pull the pin out as you rotate your hand forward to the palm-in, thumb-side up position. This twisting "cork-screw" motion makes removing the pin much easier. The grenade is now ready to be thrown. Care should be taken to avoid striking the dominant hand or arm. This may result in a dropped grenade and premature ignition. It is also a good idea to leave the pull-ring on your support-hand's index finger until after the grenade has been thrown. This way, the pin may be reinserted should the situation change and the grenade not be needed immediately. Remember to bend pull-pin tines back if you replace the pin.

Hand-Delivery Techniques

Once the pin has been pulled from the fuze, it is ready to be delivered into the objective area. That is, *the grenade* is ready to be delivered. Yes, it happens, very rarely, usually under extreme stress, that the pin is thrown while the grenade remains firmly clenched in the hand. Naturally the unwitting officer is never reminded of this embarrassing incident by his fellow officers. . . .

It is also not unheard of for a grenade to be delivered with the pin still firmly in place, having been forgotten in the heat of the moment. To ensure that these mishaps don't occur, thorough, ongoing training must be incorporated into your department's policy. A rotation stock system to help implement such a program is outlined in Chapter 19.

Overhand

The stance is wide and balanced. The pull-pin is left on the support-hand finger. The support arm is elevated and extended toward the objective area. The primary, or throwing arm, is cocked back, in a manner that can best be described as a combination baseball throw/shot-put technique.

The primary arm is pistoned forward, following the upward angle of the extended support arm. The grenade is released at the moment the primary arm is fully extended.

Underhand

The stance is wide and balanced. The support arm is extended forward toward the objective area, providing a visual index as well as helping to maintain balance. The grenade is held in the primary hand. The primary arm is brought to a rearward position, keeping the elbow straightened.

The primary arm is rotated forward from the shoulder. The grenade is released at the appropriate moment. Your accuracy in using all of these techniques depends on your training. Using dummy grenades in practice that simulate the weight and feel of the munitions that you'll deploy during a real engagement is helpful.

"Dummy" training grenades. Inert or live fuze assemblies may be used. Training grenades like these may be purchased or easily fabricated. Soft drink cans filled with sand or cement to approximate the weight of grenades may also be used.

Practicing Delivery Techniques

Practicing delivery techniques from various positions and obstacles is the best way to prepare for the tactical realities of deployment. There is no guarantee that you will always be delivering CAW grenades on flat, open surfaces. In this photo, the operator delivers grenade into the objective area using the overhand technique from an elevated obstacle. Contingency plans should be devised to address the possibility of the grenade's being unintentionally dropped while attempting a delivery from an elevated position. This is especially relevant if the operator delivering the grenade is not wearing a protective mask. Because the smoke/agent will more than likely rise, the operator may be better off getting down from the ladder or obstacle and reassessing the situation.

Training for Accuracy

Consistent and ongoing practice is necessary to develop the ability to deliver the CAW grenades with a reliable degree of accuracy. Here, dismounted rubber tires and the openings in the structure in the background are used as objective areas.

Learning through Observation

One of the best ways to learn what to expect when using pyrotechnic grenades is to observe the functioning of individually deployed units. Not only do various grenades perform differently, but the agent cloud will act differently depending on terrain and weather conditions.

Many factors, including wind direction and speed and thermal and mechanical turbulence, affect the shape and movement of the smoke/agent cloud.

Using Cover

When delivering agents into a potentially lethal environment, sound tactical doctrine must be followed. Here tactical team members use both transitory cover (the body bunker) and defensive cover (pistol- and shotgun-armed cover officers) while getting into position.

MALFUNCTIONS

Occasionally, even when the operator has done everything right, a pyrotechnic grenade will fail to operate as expected. In most cases the problem will be traced to the fuze assembly. There are several components incorporated into the common M201A1 fuze and its variants that may fail:

- Safety levers may be bent or deformed, in some cases severely enough to prevent their being released once the safety pin has been withdrawn.
- Strikers or striker springs may be damaged or deformed.
- Primers, delay elements, or igniters may fail.

Should a grenade fail for any reason, it will be necessary to recover it and render it safe before an attempt is made to determine the problem. It is a good idea to wait approximately 30 minutes before you approach or handle the device. The officer who recovers the munition should also be wearing some type of appropriate protective equipment. This equipment must be suitable to protect him from the

agent contained in the device, as well as from the heat and flames produced if the device is ignited or even explodes. A generalized procedure to accomplish this task is illustrated in the following series of photos.

Malfunctioning Pyrotechnic Munitions— Recovery Procedure

1. The first thing that must be accomplished is a visual examination of the munition. Approach it carefully and look at it from all angles. You are looking to see whether the lever has not been released for any reason. You are also looking for any signs of damage or deformity to the fuze or body. If you detect any activity whatsoever, back away from the munition and begin the 30-minute waiting period again. Paul Damery, founder of the Massachusetts State Police Bomb Squad, demonstrates this procedure with a smoke grenade that has failed to ignite.

2. If no activity is detected (e.g., smoking, sparking, "fizzing" sound) and the striker has not rotated forward on its axis, then you should secure the device with a booted foot and insert a safety pin into the fuze assembly. This will ensure that the striker does not hit the primer during handling.

3. The munition body should then be examined for signs of damage. If no damage is discovered, you can then assume that the malfunction was due to a fuze assembly failure. In these cases, you may remove the fuze by carefully unscrewing it from the grenade body (M201A1-type fuzes ONLY). The malfunctioning fuze may then be properly discarded. A new fuze may then be carefully inserted into the body, and the munition redeployed. Should the second fuze appear to operate correctly but the grenade still fails to ignite, then the munition itself must be presumed defective. The recovery procedure must then be repeated. Once recovered, the munition must then be properly disposed of in accordance with all applicable laws and regulations.

Foot approaching the trip wire.

BOOBY TRAPPING

A section on booby-trapping these devices has been included for two reasons. First, I believe that the operators who use these items should have a thorough knowledge of the many ways they may be employed. This way, should the need arise for any reason, the devices may be employed in such a manner safely and with some level of competence. Second, by being exposed to not only the means but also the very *idea* of booby trapping, the operator's awareness is expanded. This could be significant should you find yourself involved in an operation dealing with an adversary who also has the means and knowledge to use this or a similar device against you.

Unfortunately, such scenarios are not unthinkable. People involved in the illegal cultivation of marijuana have been known to booby-trap their growing fields—often on property not their own, including state and national forests. Booby traps have also been encountered more frequently in clandestine chemical laboratories. And in a tragic case that unfolded in New Hampshire in August 1997, a 67-year-old loner went on a two-state shooting rampage during which four people died (two of them state police officers) and several others were wounded. The gunman, killed during an exchange of fire, had burned down his own house during the carnage but left behind a barn filled with explosives and incendiary materials—some of which were reportedly booby-trapped.

Please understand that I am giving away no classified tricks of tradecraft or secret information in this section. The methods I outline here have been illustrated in works such as *The Anarchist Cookbook*[4] and various declassified military manuals. Using booby traps against members of law enforcement is not new; the knowledge is out there and has been for years. In fact, during past civil disturbances in this country, booby trapping was fairly common.

Firebombs, explosives, and many other kinds of devices were fabricated and used against military, police, and fire-fighting personnel in some of the more severe instances.

The widespread adoption of the personal computer and the advent of the Internet in the past few years have also facilitated a tremendous proliferation of all types of information going out to all types of people.

As strange as it may seem, though, the people who will most likely have to deal with the results of any unlawful application of this knowledge are often totally oblivious to it. That, again, is the reason this

The Trip Wire

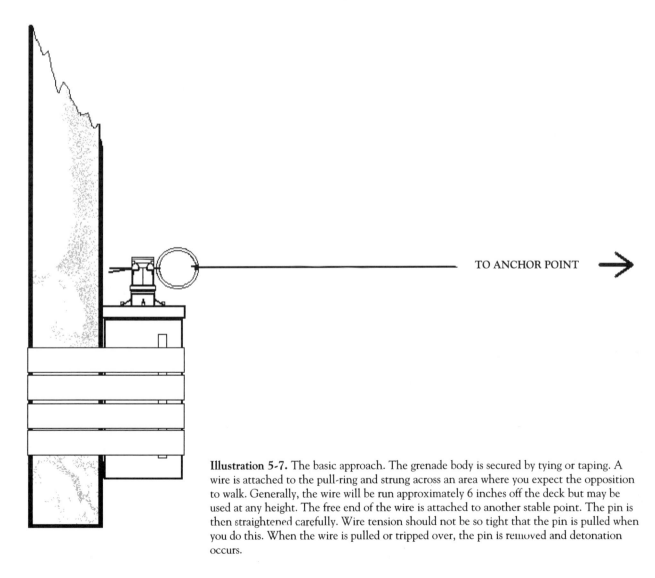

TO ANCHOR POINT →

Illustration 5-7. The basic approach. The grenade body is secured by tying or taping. A wire is attached to the pull-ring and strung across an area where you expect the opposition to walk. Generally, the wire will be run approximately 6 inches off the deck but may be used at any height. The free end of the wire is attached to another stable point. The pin is then straightened carefully. Wire tension should not be so tight that the pin is pulled when you do this. When the wire is pulled or tripped over, the pin is removed and detonation occurs.

Soup Can Surprise

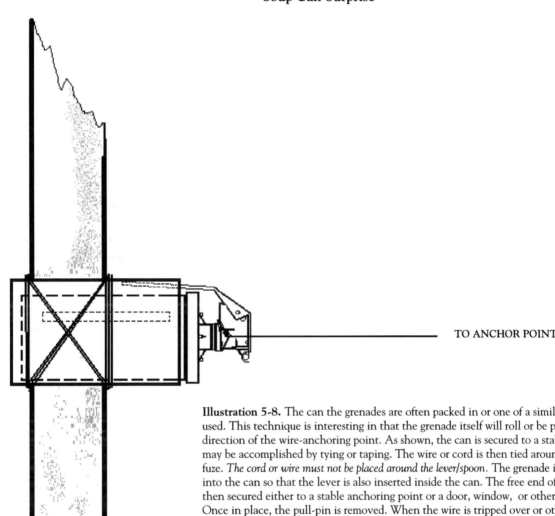

TO ANCHOR POINT →

Illustration 5-8. The can the grenades are often packed in or one of a similar size may be used. This technique is interesting in that the grenade itself will roll or be pulled toward the direction of the wire-anchoring point. As shown, the can is secured to a stable point. This may be accomplished by tying or taping. The wire or cord is then tied around the grenade's fuze. *The cord or wire must not be placed around the lever/spoon.* The grenade is then placed into the can so that the lever is also inserted inside the can. The free end of the wire/cord is then secured either to a stable anchoring point or a door, window, or other movable object. Once in place, the pull-pin is removed. When the wire is tripped over or otherwise pulled, the grenade is pulled from the can, releasing the lever and detonating the device.

Pressure-Release Gate Trap

Illustration 5-9. The grenade is placed under a gate and secured. The pull-ring is removed. When the gate is opened, the lever is released and the grenade is detonated. The lever may have to be shortened (as shown) depending on the height of the gate and the space below it. Primarily intended for use with a fragmentation-type device.

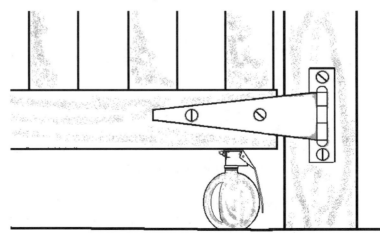

45

section is included here, for as Thomas Mann advised, "the actual enemy is the unknown."[5]

"ETERNAL VIGILANCE IS THE PRICE OF SAFETY"

Col. Rex Applegate quoted this military axiom in his book *Riot Control* while addressing the subject of booby traps.[6] The adage also applies to any operation we undertake or participate in.

We must remember that no operation is finished until it is *completely* finished. Getting sloppy or relaxing too soon is considered to be one of the 10 deadly errors in police work, but it is actually one of the more deadly errors of everyday life.

Lao Tzu addressed this reality in the *Tao Te Ching* approximately 2,500 years ago.[7] Far from being outdated or obsolete, the wisdom and advice in this work are timeless. In fact, most of us are probably familiar with several variations of the thoughts and ideas contained in the 81 chapters. For example, the quote, "A journey of a thousand miles starts under one's feet," is often expressed as "a journey of a thousand miles begins with but a single step."

While this thought may help us to achieve a proper perspective when beginning a new or difficult undertaking, the actual lesson the author was trying to convey is missed, because the entire idea was not contained in this one line. The complete thought reads:

> "People usually fail when they are on the verge of success. So give as much care to the end as to the beginning; then there will be no failure."[8]

When trying to express this philosophy while training others, I usually just sum it up as "the last step in the journey is as important as the first."

CARE AND MAINTENANCE

Grenades should be kept in their packing containers until needed. The storage area should be cool and dry and free of undue humidity.

Long-term storage of grenades in the trunks of vehicles is not recommended. There have reportedly been instances where grenades maintained in this manner have exploded when used, probably as a result of the extreme temperatures that can be easily reached in the trunk of a car on a sunny day. This is more likely to occur with CN-loaded pyrotechnic grenades as opposed to those loaded with CS because CN melts at a lower temperature than CS (129.2 to 131°F for CN, in contrast to 199.4 to 205.8°F for CS [54 to 55°C for CN; 93 to 96.5°C for CS]).

At these temperatures, the agents may melt inside the canisters and block the emission ports. Once the temperature drops, the agent becomes solid again.

Should this process be repeated, the blockage may become so severe that it prevents the escape of the gas once the grenade is ignited. The pressure inside the grenade canister may then quickly become critical, causing the body to fragment.

Pins should be left with the tines bent, not straightened, until just before use. Some officers take an additional precaution and secure the lever to the body with a piece of electrical or duct tape. *If this technique is used, the officer must be sure to remove the tape before pulling the pin.* This may seem a no-brainer, but it must be remembered that these devices are often used under highly stressful conditions. Again, there have been cases where grenades (with pins removed) have been introduced into an objective area only to not be detonated because tape was left wrapped around the lever and body. This is why some officers consider taping the lever down as not only unnecessary, but a violation of the KISS principle. Personally, I believe the extra precaution is well worth taking. This opinion was reinforced a few years back by an experience one of my teammates had with a distraction device.[9]

As members of a statewide response team, we were required by logistics to have all of our equipment with us 24 hours a day. So we had to keep our gear stored in the trunks of our individually assigned cruisers. During the course of the year, this equipment would be removed and repacked frequently. Often, the need to deploy was immediate and the operational environments were unfriendly— which meant that our gear took a beating.

One evening, while responding to a situation of an armed and barricaded suspect, one team member removed some gear from his load-bearing tactical body armor while suiting up. The unsettling sound of a spoon flying off caused him to immediately push another officer out of the way just as the flashbang ignited, producing a flash of approximately 3.5 million candlepower and a bang registering about 170 decibels. Both officers escaped critical injury, though my teammate did suffer some minor nerve damage to his hand caused by the extreme pressure wave radiated from the reflecting surface of the vehicle's interior. It was later determined that the fuze base had been damaged and the safety pin inadvertently disengaged as a result.

Could a similar situation occur with any other type of pull-pin device? Of course. While the results would not be so stunningly dramatic should this occur with a pyrotechnic or aerosol grenade, they could be just as devastating, if not more so.

The possible nightmare scenarios of an unintentional or premature detonation are endless:

- Any type of flame-producing pyrotechnic grenade set off in the trunk or interior of a vehicle, or while being held by, or secured on, an officer's person
- Any type of chemical agent or smoke grenade being detonated in the outer perimeter or just prior to the execution of a tactical intervention

The levers on these canister grenades have been secured with lengths of tape for reasons discussed in this chapter. If you decide to use this precaution, the running end of the tape should be folded over about 3/4-inch for easy removal.

- Any type of unintentional detonation in an area where friendlies (e.g., other police officers, noninvolved personnel, elderly citizens, children) are present

Reality demands that we put ego aside and admit that these types of incidents can and do happen in the real world. Putting our occasionally substantial egos aside is also the first step toward minimizing the chances of such incidents. Other steps—such as conducting frequent, viable training operations and performing regularly scheduled inspections and maintenance of equipment—must also be integrated into the program.

As I was advised long ago by a gentleman who had "seen the elephant" many times, "Never take for granted things that can only go wrong once."

NOTE: For additional information on storage areas, maintenance, and rotation of stock, see Chapter 19.

NOTES

1. One pro-rioter Web site advises that "wrist-rocket" slingshots should be used to launch marbles and ball bearings at the police during civil disturbance situations.
2. Oxygen levels below 19.5 percent are considered life threatening.
3. A plastic "friction fuze" is used on some Brazilian-made devices but is not commonly employed in the United States.
4. William Powell, *The Anarchist Cookbook* (New York: Lyle Stuart, 1971). First published in 1971, this classic is still around and still interesting reading, if for no other reason than to understand the mentality of the times that produced many of the "leaders" we suffer today.
5. The Nobel Prize–winning German novelist and critic considered to be one of the most important figures in early 20th-century literature. In 1933, Mann chose self-exile from his homeland rather than living under Hitler's rule.
6. The revised edition of *Riot Control* was published in 1981 by Paladin Press but is now out of print. It is a must-have for the serious student of the tactical police sciences.
7. Lao Tzu, depending on the historian writing about him, was either an older contemporary of Confucius who lived in China from 551–479 B.C. or, conversely, not an actual historical figure at all. Many historians believe that the name Lao Tzu was simply attributed to a collection of anthologies compiled from short passages during the "golden age of Chinese thought."
8. *Lao Tzu. Tao Te Ching: A New Translation*, by Gia-Fu Feng and Jane English (New York: Vintage Books, 1972), ch. 64.
9. Distraction devices, or "flashbangs," while not true grenades (they do not emit metal fragments, gas, smoke, or stingballs) are generally configured similarly in that they have a cylindrical body and a pull-pin/safety lever fuze assembly.

Instantaneous-Blast Grenade

The versatile multipurpose grenade (MPG).

HOW THEY WORK

Described as instantaneous-blast, blast-dispersion, or expulsion grenades, these devices use no fuel, flames, or heat to disperse the agent. Rather, micropulverized agent particles are mixed with a buffering agent that keeps the load in powdered form. Upon detonation, the agent is released practically instantaneously into the atmosphere. These types of grenades are generally produced to function in one of three ways, as an explosive charge, compressed gas, or propellant-driven piston.

Explosive Charge

A small explosive charge surrounded by micropulverized agent is detonated, expelling the agent into the atmosphere through ports in the grenade body. Some grenades of this type also employ perforated or scored cardboard, plastic, or aluminum bodies that burst when the charge is detonated, releasing the agent as well as possibly producing fragmentation. It is critical that any fragmentation-producing grenades be deployed as safely as possible. They must never be thrown directly at an individual, particularly near the area of the face. There is also a danger of these devices' going off in the operator's hand and causing injury.

A blast-dispersion grenade after detonation. The aluminum body is scored along the sides to allow it to burst with a minimal chance of producing fragmentation.

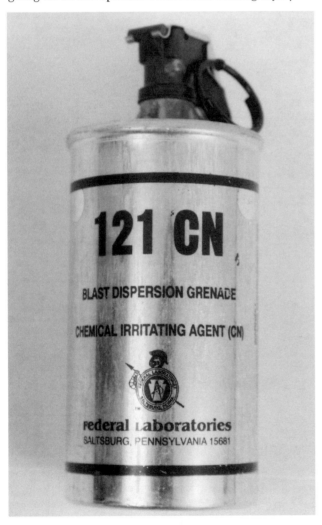

Federal Laboratories' 121CN Blast Dispersion Grenade. The 514CS grenade is similar except it is loaded out with CS. The 121CN holds approximately 44 grams (1.54 ounces) of dry-powder agent.

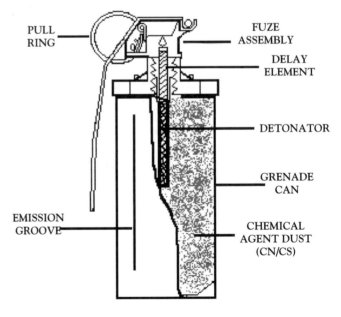

Illustration 6-1. The 121CN Blast Dispersion Grenade uses an M201A1 fuze with a delay time of approximately 1 to 2 seconds, making launching impractical.

The grenade in the center is an obsolete M25A2 military bursting grenade. This grenade is loaded with CS dust agent. The previous model, the M25A1, was loaded with CN, but functioned similarly. The hard-plastic grenade body has a diameter of approximately 3 inches. When the pin is pulled, you have approximately 1.5 to 3.0 seconds to deliver it to the objective area before it detonates, emitting an agent cloud of approximately 5 meters (16.0 feet) in diameter and fragmentation up to 25 meters (81 feet). These grenades also tend to explode when dropped on hard surfaces even if the pin has not been pulled. Even though this grenade is not recommended for use, I have included it here because it or similar devices (such as the Penguin baseball-type grenade CN/X5) may be stored away in some police departments' chemical agent inventories. This one was located in such an inventory along with rubber pyrotechnic "Han-ball"-type grenades. Confusion between these types of devices could prove tragic. This is one of the reasons that chemical agent weapon inventories must be monitored and maintained, and expended (Model 119, pictured top right) obsolete, or expired munitions properly disposed of.

Micropulverized chemical agent is forced out of the grenade body and into the atmosphere by the release of compressed CO_2 gas.

This gas, contained in a metal cartridge similar to those used in CO_2-powered pellet guns, is released after the cartridge has been pierced. The piercing is accomplished by using a standard M201A1 fuze and a small wadded charge of black powder. After the safety pin is pulled and the lever released, the fuze detonates the black-powder charge, forcing the CO_2 cartridge inside the grenade body against a fixed pin. The gas is then released, and the agent is forced out through an emission port (or ports). The agent may be forced out of the grenade body either directly by the pressure of

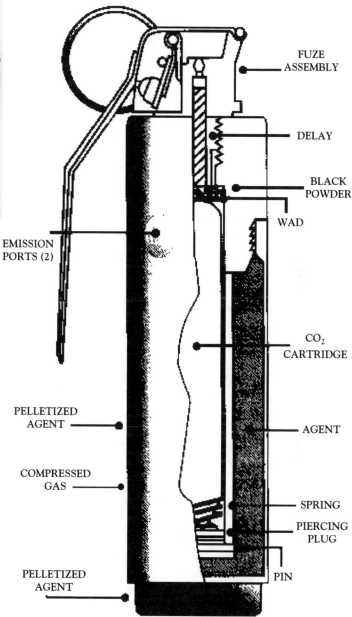

Illustration 6-2. Defense Technology's T-16 grenade is not suitable for launching or airdropping because of its instantaneous-discharge feature. (Courtesy of Defense Technology.)

51

the gas or by way of a piston driven forward by the pressure of the gas.

The T-16 (shown in Illustration 6-2) is approximately 7 1/2 inches (19 centimeters) long and 1 1/2 inches (3.8 centimeters) in diameter. Like the company's F-16 model that it replaced, the T-16 uses a CO_2 gas cartridge to expel the micropulverized chemical agent into the atmosphere. The grenade is equipped with a standard M201A1 fuze set for a 1-second delay. After the safety pin is pulled and the lever released, the fuze burns down and then detonates a small, wadded black-powder charge, which forces the cartridge inside the grenade body against a fixed pin. The gas is released when the pin pierces the cartridge body, and the agent is forced out through two emission ports located on the sides of the grenade body. The device, containing approximately 9 grams (0.32 ounce) of CN or CS agent—0.70 grams (0.025 ounce) when loaded with OC—will fill a 2,000-cubic-foot room (10 x 10 x 20 feet) in about 5 seconds.

Propellant-Driven Piston

A propellant-driven piston is used to expel finely ground particles of the agent directly into the atmosphere. This design is most notably embodied in the unique multipurpose grenade (MPG) produced by Federal Laboratories.

THE MULTIPURPOSE GRENADE

The Fuze

The MPG uses a unique variable-delay mechanical fuze, which is inserted into the body of the grenade as a unit. A time-selector lever on top of the unit is keyed to the delay tube and can be set for a delay of either 2 or 5 seconds.

When the safety pin is removed, a spring-loaded striker causes the safety lever to fly off, exposing a primer. The striker continues forward and hits the primer, igniting the selected delay element. Once the element is burned through, a propellant gas is released in a specially sealed chamber. The expanding gas then forces a mechanical piston forward toward the base of the grenade. The micro-pulverized chemical agent is

The unique MPG, designed according to federal specifications, was first developed in the early 1960s. Note the special rifling band on the base of the grenade. When fired from the rifled MPG launcher, the grenade spirals like a bullet. This allows for a more stable flight trajectory and overall accurate placement of the munition when launched.

A time-selector lever on top of the unit is keyed to the delay tube and can be set for a delay of either 2 or 5 seconds. If left at the center position, the grenade will not detonate when the lever is released!

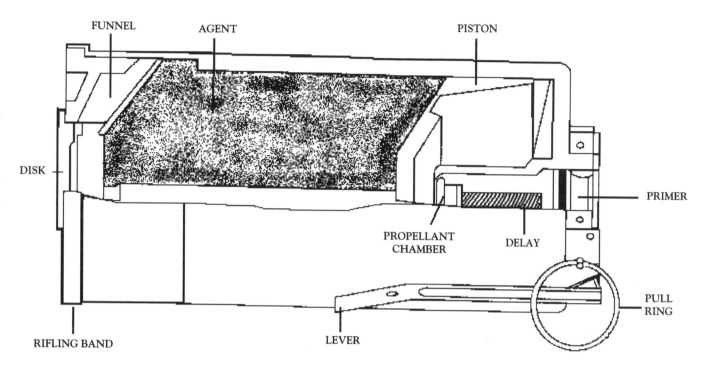

FUNNEL   AGENT   PISTON   DISK   PRIMER   PROPELLANT CHAMBER   DELAY   PULL RING   RIFLING BAND   LEVER

Illustration 6-3. The parts of an MPG.

then projected in the form of a cone-shaped gas cloud for about 20 feet (6 meters).

A special disk that covers the exit port at the base of the grenade is ruptured by the force of the blast and falls to the ground at low velocity. If released downwind (optimal wind speed of 3–5 mph), the cloud will rapidly expand, becoming invisible as the fine gas particles float downrange. Given optimal conditions, the cloud's effects may be felt as far as 200 yards (about 183 meters) from the release point.

Another unique feature of the MPG is its ability to deploy its payload while the device is held in the hand. This proves useful when dealing with crowds or pockets of agitators. The grenade may be maneuvered into the exact location needed to disperse the agent for optimal effect. When using the grenade this way, a slightly different grip than normally used is required.[1] The grenade, fuze set (generally to 2 seconds), may be held in either hand, though the use of the support hand is recommended because this allows the user to access the handgun

An MPG being detonated while using a static-hand technique. Note the cone-shaped cloud. Under optimal conditions this cloud will be ejected approximately 20 feet into the objective area. The low velocity of the ejected agent allows it to be aimed in the direction of people without fear of causing injury. Once airborne, the agent will continue to travel downwind, its effects being felt as far as 200 yards downrange.

with the primary hand if necessary. The user should hold the lever down by the thumb, while maintaining a solid grip on the unit's body. The exit port must be pointed toward the objective area. It is a good idea to aim the port slightly upward (as much as 30 degrees) to ensure good dispersion of the agent. The MPG is very user friendly. The grenade, when detonated, produces no concussion or recoil. There is also no heat or shrapnel, and noise is negligible.

OTHER MPG DELIVERY METHODS

Launching the MPG

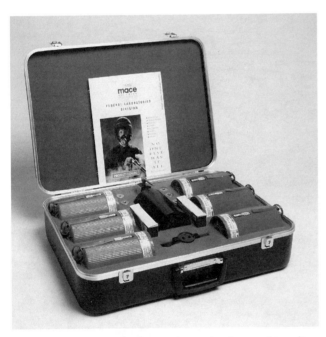

MPG emergency kit. This kit was designed to be stored in police cruiser trunks or emergency response vehicles. It holds six MPGs, shotgun launching adapter and launching cartridges, and 12-gauge barricade-penetrating rounds.

The MPG may also be launched to distances of more than 100 yards (approximately 90 meters) using a special adapter with a 12-gauge shotgun. To launch the MPG, the grenade is first placed into the adapter, base first, the safety lever being secured inside the launching cup. The timer must then be set for 2 or 5 seconds, depending on the desired effect (air burst or ground burst). The pin is then removed. Once the pin is removed, the steel safety lever will exert pressure against the side of the launching cup.

This pressure is usually enough to keep the MPG securely in the launching cup until fired. The blank launching cartridge should be loaded last for safety's sake. The MPG is then ready to be launched. Because of the effect of the rifling band, the MPG spirals through the air, providing it with a more accurate and consistent trajectory than found when launching normally configured grenades. This allows it to be introduced into objective areas through windows and other openings out to distances of 75 yards (69 meters). The grenade may also be fired directly at a hard surface (such as a wall above the desired objective area) if needed. This will cause the grenade hull to break, releasing the agent.

Air-Dropping the MPG

The MPG may be airdropped onto buildings or into crowds fairly safely because it produces no heat, flames, fragments, sparks, or concussion. The plastic grenade body also has no sharp edges, points, or fins, and at 1/2 pound (0.19 kilogram) empty is surprisingly light.

Booby Trapping

The MPG can be booby-trapped using techniques similar to those shown in Chapter 5, though they may not be as efficient because of the MPG's single-port, instantaneous-discharge design. There is, however, another booby-trapping method unique to the MPG that may be used. An electrically activated screw-in socket connection may be used to replace the variable-delay fuze. The MPG can then be screwed directly into a standard light socket. When the switch is turned on or otherwise activated, the grenade disperses its payload. This system is reportedly used at U.S. government and embassy installations.

INSTANTANEOUS-BLAST GRENADES— STRENGTHS AND WEAKNESSES

These devices can be used outdoors and are generally considered appropriate for indoor use as well, since they produce no flames or extreme heat. The instantaneous-blast, blast-dispersion, or expulsion grenades disperse their powder or dust-like agent payload in approximately 1/10 of a second

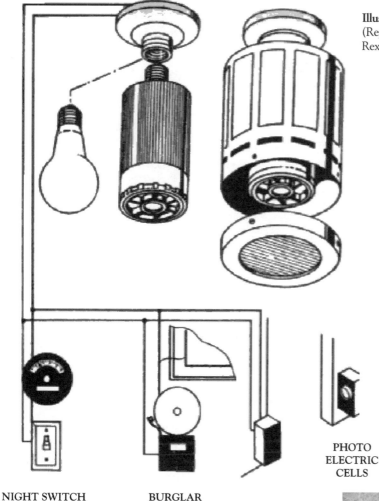

Illustration 6-4. Electronically activated MPG configuration. (Reprinted from Riot Control: Materiel and Techniques by Col. Rex Applegate.)

PHOTO
ELECTRIC
CELLS

NIGHT SWITCH
OR TIMER

BURGLAR
ALARM SYSTEM

upon detonation. This almost instantaneous method of dissemination is advantageous because it produces a more persistent irritating effect than pyrotechnic grenades, while practically eliminating the dangers of fire, injury from fragmentation, and throwback.

One of the major disadvantages with these types of devices is that with the exception of the MPG, they tend to not expel their entire payload when detonated. This occurs because at the moment of the blast, not all the agent is expelled through the constricted emission ports or grooves.

CARE AND MAINTENANCE

Grenade shelf life will vary with device and manufacturer. All expiration dates should be noted,

Blast-dispersion grenades after detonation. Note the amount of agent powder directly around the grenades. This occurs because at the moment of the blast some of the agent is not expelled from the canister through the emission ports or grooves.

and stocks should be rotated as outlined in Chapter 19.

The MPG, being a sealed unit with no taped-over gas ports or porous components, may be stored safely for a longer period than other standard grenade munitions. Based upon optimal storage conditions, the MPG may be maintained for up to six years.

Care should be taken not to strike the exit port cover blow-off disks, because they may occasionally come loose.

MALFUNCTIONING INSTANTANEOUS-BLAST GRENADES—RECOVERY PROCEDURE

Should an instantaneous-blast munition fail, a protocol similar to that used for pyrotechnic devices should be employed. A 30-minute waiting period should be observed and proper protective equipment used. The device should then be approached carefully and visually examined. If the cause of the failure is not apparent, then the device should be recovered and properly disposed of.

NOTE

1. The MPG can be deployed using any of the hand-delivery techniques shown in Chapter 5. Due to its light weight, the MPG may also be introduced into an objective area by throwing it like a baseball. When using an overhand, underhand, or baseball throw, the grenade should be held securely with the safety lever in the web of the hand.

Aerosol Grenades

Aerosol grenades offer another viable option to law enforcement and military tactical operations personnel. In effect, the aerosol grenade is basically a hybrid of standard discharge-type grenade munitions and hand-held ASR sprays.

These grenades tend to disperse their entire payload in a matter of seconds, negating the possibility of a throwback. Most of these devices use non-flammable solvents in pressurized containers to deliver the agent in a quickly spreading cloud. Aerosol grenades can be configured like standard CAW grenades with M201A1-type fuzes or industry-standard aerosol cans as used by Aerko International.

NON-FUZED AEROSOL CANISTER GRENADE

The munitions shown below are loaded with a combination of CS and OC that is suspended in a nontoxic, nonflammable carrier agent or "vehicle." An ultraviolet (UV) marking agent is also included in the formulation. Once released, the gas, chemical agent, and UV marker agent are for all intents and purposes invisible. If eye protection is worn, vision will not be obscured by the resulting mist as when using pyrotechnic munitions that produce dense clouds of billowing smoke. This has its advantages, especially when tactical entry needs to be made into an enclosed environment after CAWs have been introduced.

The contents of the can are pressurized by what the manufacturer claims to be a "special environmentally safe gas-mixture which is engineered so as to maintain a constant pressure in the can until the entire contents are expelled."

The agent itself, classified as an irritant, is effective. I have experienced its effects under both training and operational field conditions. The results are what you would expect from a combination of CS and OC: burning and tearing of the eyes, nasal discharge, and uncontrollable sneezing. Because the agent is suspended in a mist, there is also a greater tendency to inhale or ingest it. Other effects, such as a burning sensation in the bronchial tract and lungs, are then felt, making it difficult to breathe. This in

AERKO International's CLEAR OUT aerosol grenade contains 1 percent CS, 1 percent OC, and ultraviolet marking dye.

turn often produces feelings of panic or disorientation in those exposed to the agent.

As alluded to above, the aerosol grenade is also very useful for indoor applications. Besides the fact that there is no risk of explosion or fire hazard, the agent evaporates fairly quickly under normal circumstances (40 to 50 minutes) and leaves no perceptible residue.

How It Works

AERKO International's CLEAR OUT grenade, unlike a pyrotechnic or instantaneous-discharge grenade, is loaded into an industry-standard aerosol can. This familiar package comes complete with a plastic protective cap.

No percussion, mechanical, or electrical fuzes are used; rather, a molded plastic actuator valve on top of the can is equipped with a "snap-down" lever, which locks the valve into the open configuration when it is depressed. Once depressed, the valve allows the pressurized contents to escape, emptying the 6 fluid

CLEAR OUT aerosol grenade specially configured with "auto-ejector" accessory. This allows the agent to be administered directly into the interior of a vehicle without having to break any windows. The needle is passed through the rubber gasket that surrounds the door or window. Press the actuator for 2 to 3 seconds and the agent will enter the vehicle and immediately vaporize and expand, filling the entire compartment.

Holding the Aerosol Canister Grenade

CLEAR OUT aerosol grenade specially configured with "keyholer" accessory. This allows the agent to be administered through small access holes or under doors. Care must be taken when employing the devices in this manner. If the tubing is not positioned and secured properly, personnel other than those intended may be affected. The author was present when this happened during an actual operation. He was not amused. The way to avoid this type of mishap is through proper and continual training.

After removing the protective cap, grasp the can, actuator up, as shown. The actuator valve should be facing away from you. The thumb is used to depress the lever. Once the lever is depressed, the contents will immediately begin to be discharged.

Deploying the Aerosol Canister Grenad

The grenade, once activated, can be deployed using any of the techniquesas illustrated in Chapter 5. It may also be rolled into the objective area, or held in the hand as it discharges (shown above). A stationary chemical barrier may also be established by forming a line of discharging cans in the upright position. NOTE: This munition is not launchable.

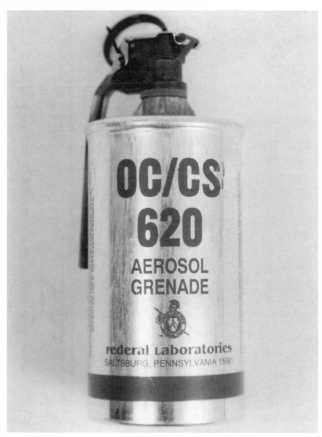

Federal Laboratories' 620 OC/CS aerosol grenade.

ounces (170 grams) in approximately 28 seconds.[1]

One can of CLEAR OUT will reportedly affect an area of approximately 23,000 cubic feet.[2]

FUZED AEROSOL GRENADE

An aerosol grenade can also be configured to resemble and function similarly to a standard CAW grenade. Federal Laboratories' 620 OC/CS Aerosol Grenade is loaded out with a small aerosol canister containing a combination of OC (6-percent concentration)[3] and CS suspended in a nontoxic, nonflammable carrier agent. When grenade is detonated, 100 percent of the contents is discharged.

How It Works

Approximately 1 to 2 seconds after the M201A1 fuze has been activated, a small charge is detonated

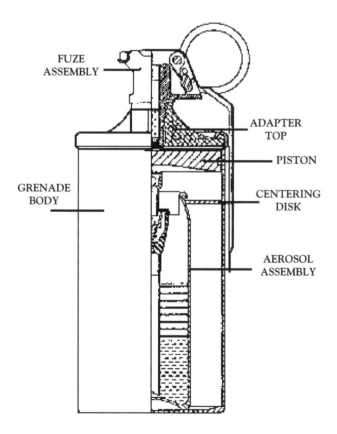

Illustration 7-1. The 620 OC/CS Aerosol Grenade. (Courtesy of Federal Laboratories.)

inside the grenade body. The force of this charge drives a piston forward, releasing the pressurized aerosolized agent through emission ports in the grenade body. Within 3 seconds of this activation, 100 percent of the aerosolized agent is released.

AEROSOL GRENADES— STRENGTHS AND WEAKNESSES

Aerosol grenades are obviously ideal for indoor use. They discharge 100 percent of their payloads and, depending on which type is used, will do this in a time span of 3 to 28 seconds. The use of these devices also eliminates the hazards associated with other types of munitions such as fire or injury by concussion or fragmentation.

The configuration of the CLEAR OUT grenades also allows the devices to be employed in various ways by using the different accessories available.

MALFUNCTIONING AEROSOL GRENADE MUNITIONS—RECOVERY PROCEDURE

Should an aerosol grenade munition fail, it should be approached carefully and visually

examined. If the cause of the failure is not apparent, then the device should be recovered and properly disposed of. Proper protective equipment should be used during the recovery and disposal processes.

CARE AND MAINTENANCE

It is recommended that CLEAR OUT aerosol grenades be stored at temperatures under 120°F (48.8°C). The manufacturer also advises that prolonged exposure to temperatures below 32°F (0°C) will result in a slower discharge rate. The cans should be kept dry because they tend to rust. They must also be protected from being punctured.

The aluminum-bodied Federal Laboratories 620 grenades should be stored in their packing containers until needed. The storage area should be cool and dry and free of excessive humidity.

NOTES

1. This is a time estimate based on an ambient temperature of 70°F. Discharge time will increase at significantly lower temperatures.
2. For example, a building 50 x 50 feet with an 8-foot ceiling contains 20,000 cubic feet.
3. 1,000,000 Scoville heat units.

Pyrotechnic Projectiles

Pyrotechnic projectiles, ranging in size from 6 inches (15 centimeters) to 12 inches (30 centimeters), are self-contained rounds of ammunition. They are designed to be fired from gas launchers produced expressly for this purpose (see Chapter 14). No launching adapters or special launching cartridges are needed. The short-range projectiles are intended for limited-distance applications. Smoke projectiles can be used for training, signaling, or verifying wind direction prior to deployment of chemical agents. These munitions may contain single or multiple individual chemical loads, and their primary purpose is to aid in routing or dispersing unruly crowds in civil disturbances.

As with pyrotechnic grenades, these projectiles are not intended for indoor use since both possess fire-producing capabilities. In extreme situations, however, special barricade-penetrating rounds may be needed to penetrate windows, doors, walls, or siding. The decision to employ these types of rounds against personnel who have taken up barricaded positions within a structure must be made by the incident commander. This decision cannot be made lightly.

Before these munitions are used, the commander must weigh the immediate tactical needs of the situation against the potential for inducing collateral damage to any personnel—suspect, hostage, or bystander—inside *or behind* the objective area. This collateral damage may take the form of direct injury to parties struck by the fast-moving projectiles or by injuries or damage that may result from any fires the rounds initiate.

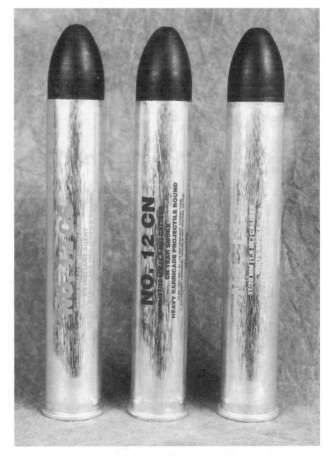

Pyrotechnic projectiles. Generally produced in 37/38, or 40mm diameters, these munitions may be loaded with chemical agents or smoke. Projectile bodies are most often made of aluminum. None of these rounds should be fired directly at any personnel: serious injury or death may result! (Shown above, left to right: DEF-TEC No. 12 CS, CN, and smoke heavy barricade rounds. These rounds are equipped with a 3-ounce [90-gram] lead tip. Overpenetration must be considered before using.)

Once the decision has been made to employ pyrotechnic projectiles against personnel inside a structure, the munitions must be used in a correct and efficient manner. Reasonable safety considerations for all involved personnel must also be addressed (see Chapters 16 and 17).

HOW THEY WORK

Rounds are produced in various configurations. All the pyrotechnic rounds illustrated in this section, however, are designed to be fired in a similar fashion.

Once properly loaded into the appropriate gas launcher, the round is fired by depressing the launcher's trigger. A firing pin strikes a primer on the base of the shell casing. This in turn initiates a wadded black-powder charge that propels the inner projectile out of the shell casing and into the smooth or rifled bore of the launcher. The round then exits the muzzle of the launcher and travels downrange. Both the maximum range and the maximum effective range vary according to the specific munition used.

This series of photos shows 37mm round being loaded and fired. (Photos by Al Pereira.)

Fin-Stabilized Rounds

Fin-stabilized models appear and perform like minimissiles. They are usually used to deliver a weighted, barricade-penetrating projectile accurately to a specific spot or area within an objective area. Approximately 1 second after the round has been fired, the smoke or chemical agent payload begins to burn. The fin-stabilized trajectory allows for accurate placement of the round into areas no larger than a standard door or window from ranges up to 100 yards (91.4 meters). Shown above is an obsolete Federal Laboratories' 530CS FLITE-RITE 37/38mm munition. The fins are spring loaded, lying flat against the body of the inner projectile while it is encased within the outer shell. Once fired, the fins open and lock into place, providing stability to the projectile's flight path (coin used to illustrate scale). Note: These munitions have been replaced by the Federal Laboratories' heavy-duty barricade-penetrator rounds.

Unstabilized Rounds

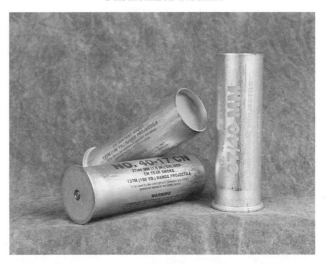

Unstabilized pyrotechnic rounds dispense smoke or chemical agents by rapidly burning their payloads. They are best suited for delivering their payloads outdoors for the purposes of dispersing or routing rioting individuals or crowds during civil disturbances. These rounds, useful out to ranges from 75 to 150 yards (68.6 to 137.1 meters), are not designed to penetrate barricades. Shown in this photo are 37/38mm Defense Technology No. 40-17 Long-Range Projectiles.

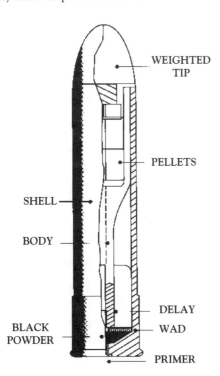

Illustration 8-1. No. 12 heavy barricade round. This 37/38mm fin-stabilized round with weighted tip begins to function by rapid burning of the payload after a 1-second delay. (Courtesy of Defense Technology.)

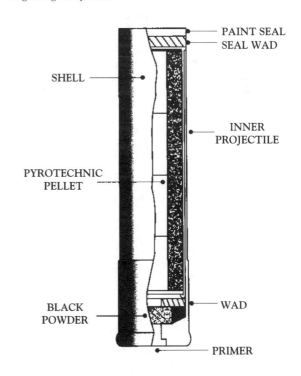

Illustration 8-2. No. 17 long- and short-range rounds—37/38mm CN, CS, and smoke. This unstabilized round dispenses a chemical agent or smoke by rapid burning in 20 to 30 seconds. (Courtesy of Defense Technology.)

Unstabilized Submunition Rounds

In this photo, multiple airborne rounds can be seen en route to the objective area. Pyrotechnic rounds of this type are designed to distribute multiple submunitions with one shot. In this way, greater outdoor area coverage may be achieved. When employing multiple submunition pyrotechnic rounds, it is imperative that the fire hazard potential be taken into consideration. These munitions also generally have less of an effective range than single projectile devices.

MALFUNCTIONING PYROTECHNIC PROJECTILE MUNITIONS—RECOVERY PROCEDURE

Should a pyrotechnic projectile munition fail to fire after the gas launcher's trigger has been depressed and contact between the firing pin and primer, a second attempt should be made. If the munition fails to fire in the second attempt, then it should be carefully removed from the launcher and properly disposed of. *No further action should be taken!*

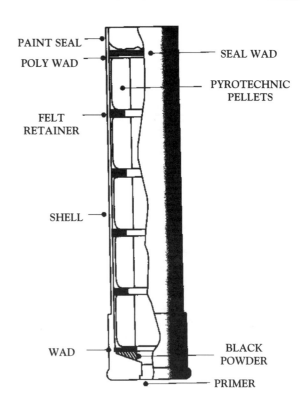

Illustration 8-3. No. 19 submunition round—37/38mm CN, CS, and smoke. This pyrotechnic round contains five separate submunitions that function individually when the round is discharged. This munition is used to disperse chemical agents or smoke over a wide area. (Courtesy of Defense Technology.)

Should a pyrotechnic projectile round be successfully launched, yet fail to ignite its chemical payload, then a similar recovery procedure to that used for pyrotechnic grenade devices should be employed. A 30-minute waiting period should be observed, and proper protective equipment should be used. The device should then be approached carefully and visually examined. If no activity is detected, then the device should be recovered and properly disposed of.

CARE AND MAINTENANCE

As with pyrotechnic grenades, pyrotechnic projectiles should not be stored in the trunks of vehicles. They must also be protected from high temperatures, humidity, and moisture. The manufacturer's recommended shelf life must also be taken into consideration.

Should a pyrotechnic projectile munition fail to fire after the second attempt, then it should be carefully removed from the launcher and properly disposed of. If possible, you should wait at least 1 minute after the last attempt to fire before opening the breech and removing the round. This is a precaution against any possibility of a "hang fire." (Photo by Al Pereira.)

Nonpyrotechnic Projectiles

Nonpyrotechnic projectiles allow for the introduction of chemical agents into a house, building, or vehicle without incurring the high risk of fire that is associated with the use of pyrotechnic projectiles. These munitions may also be employed from a safer distance than hand-held munitions.

Nonpyrotechnic projectiles are available in 37, 38, or 40mm diameters for use in gas launchers, and 12-gauge rounds that may be fired through standard, unmodified 12-gauge shotguns are also available.

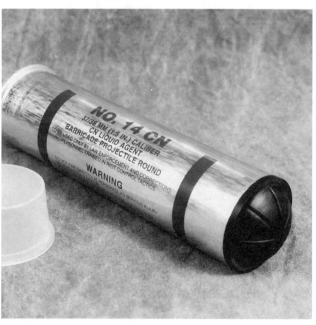

No. 14 CN 37/38 mm CN Liquid Agent Barricade Projectile Round.

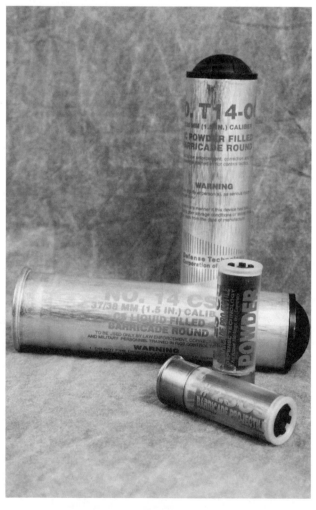

Nonpyrotechnic projectiles. These munitions are commonly referred to as "ferret rounds," though the "Ferret" designation was originally registered by Federal Laboratories to describe its own product.

HOW THEY WORK

These munitions are primarily designed to penetrate barricades. Like the pyrotechnic barricade-penetrating projectiles, these are encased in a center-fire, primer-equipped outer shell casing. They also have fins for in-flight stabilization.

Unlike pyrotechnic munitions, however, these rounds do not use combustion to disperse the payload. Rather, the hollow plastic body of the projectile is filled with CN, CS, OC, or a combination of these in a liquid or powder state.

The round, loaded into a gas launcher or shotgun, is fired when the trigger is pressed and the firing pin strikes the primer at the base of the casing. A smokeless-powder charge is then ignited and the round forced out of the casing and into the bore of the launcher/shotgun barrel by the expanding gas that results from the fast-burning powder. Plastic and foam wadding inserted into the round between the projectile and the propellant protect the projectile from the burning gas.

After exiting the muzzle, the projectile flies through the air in a stable trajectory. After it strikes a hard object, the projectile disintegrates and disseminates the chemical agent payload in either a fine mist form (liquid load) or cloud (powder load) throughout the objective area.

Liquid-/Powder-Filled Barricade Rounds

The 12-gauge rounds shown here travel at an average velocity of 1,000 fps and can disable subjects in enclosures of up to 1,000 cubic feet. The rounds penetrate most car and truck side windows, though they might be deflected off windshields and rear windows due to their angle of installation or set. Extremely useful for barricade situations, these rounds will penetrate a hollow-core door and double-panel storm windows and screen at normal tactical distances (within 100 yards, or 91.4 meters). They are highly accurate. With practice, you should be able to hit within a 1-foot radial group at approximately 50 yards (45.7 meters).

Upon striking a solid object, the projectile disintegrates, dispersing a concentration of CS, CN, or OC in a fine mist or powder. As with the pyrotechnic projectiles, nonpyrotechnic projectiles can cause severe injury or death should they strike a human being.

37/38/40mm Barricade Rounds

Loaded with enough irritant to disable subjects in enclosures of up to 4,500 cubic feet (127.4 cubic meters), the 37mm and 40mm ferret rounds are also accurate. The 37mm round will hit within a 1-foot (.3-meter) radial group at 50 yards (45.7 meters) fairly consistently. The 40mm round will hit within a 1-yard (.91-meter) radial group at approximately 110 yards (100.6 meters). The 40mm round will also

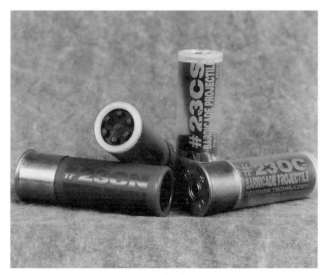

Defense Technology's No. 23 CS Barricade Projectile.

The face of this 12-gauge projectile has been designed to allow it to achieve more "purchase" on the striking surface. This is incorporated to help reduce the chance of the round ricocheting off the objective area surface.

This side view of the 12-gauge, nonpyrotechnic, liquid-load projectile allows you to see the general shape of the fin-stabilized round inside the casing.

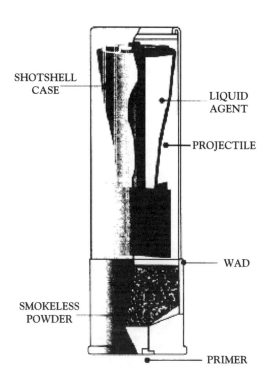

SHOTSHELL CASE

LIQUID AGENT

PROJECTILE

WAD

SMOKELESS POWDER

PRIMER

Illustration 9-1. No. 23 Liquid Filled Round. This 12-gauge fin-stabilized round, similar to the Federal Laboratories' Ferret, is designed to break upon impact, disseminating the chemical agent in mist form throughout the objective area. The No. 23 also contains a red marking dye to provide visual reference to assist with shot placement. Other rounds such as the Defense Technology T-23 use powdered agents that disseminate the agent in a cloud. (Courtesy of Defense Technology.)

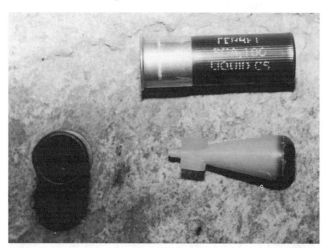

This side view of Federal Laboratories' 12-gauge Ferret round. Depending on the circumstances, 12-gauge Ferrets may penetrate a windshield out to 100 feet (30.5 meters), 3/4-inch plywood at 75 feet (23.0 meters), 1/4-inch plate glass at 200 feet (61 meters), a hollow-core door at 175 feet (53.3 meters), and a double-panel storm window and screen at 200 feet (61 meters).

These officers are firing 12-gauge, nonpyrotechnic, liquid-load projectiles into the objective area during a training exercise.

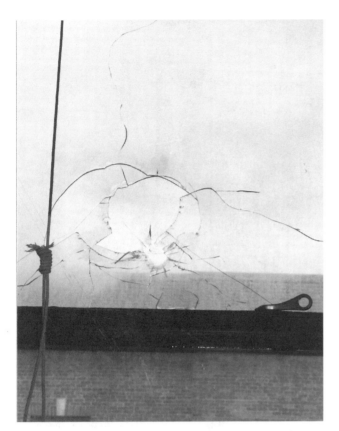

Evidence of where a 12-gauge, nonpyrotechnic, liquid-load round penetrated two panels of 1/4-inch glass from approximately 30 yards (27.4 meters) at an upward angle. All rounds of this type should be aimed to impact at or near the ceiling of the objective area to ensure that their impact is on a solid surface, breaking and disseminating the agent in an efficient pattern. Note the marking-dye stain around entrance.

The front of this recovered projectile body indicates that it disintegrated upon impact, thus releasing its agent in a fine mist form. It is imperative that these rounds strike semihard surfaces to ensure that the body fractures.

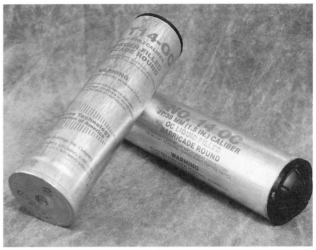

These 37/38/40mm liquid- and powder-filled barricade rounds are designed to be fired through gas launchers in the appropriate caliber. The Defense Technology rounds shown above travel at approximately 400 feet (122 meters) per second.

The projectile's impact surface areas are configured to allow them to strike and disintegrate upon impact.

Federal Laboratories' 37/38mm Model SGA 300 Ferret. This cutaway view allows you to see the design. The fins provide for in-flight stabilization. The liquid or powder load is held in the hollow core of the body. The head is designed to fracture or disintegrate upon impact with a solid surface.

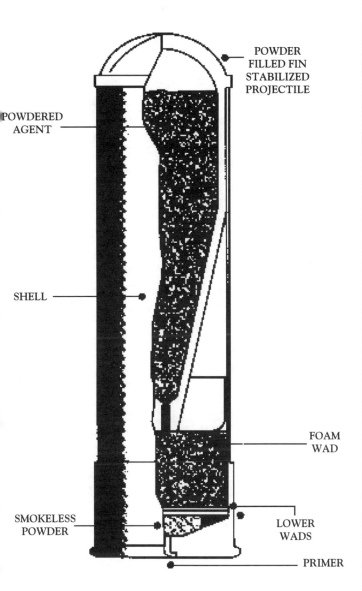

POWDER
FILLED FIN
STABILIZED
PROJECTILE

POWDERED
AGENT

SHELL

FOAM
WAD

SMOKELESS
POWDER

LOWER
WADS

PRIMER

Illustration 9-2. Defense Technology T-14 Dry Powder-Filled Barricade Round. This 37/38mm fin-stabilized round is designed to break upon impact, disseminating the chemical agent in cloud form throughout the objective area. The agent will be most effective immediately after dispersal. The time it takes for the effects to diminish depends in great part on airflow. (Courtesy of Defense Technology.)

Officers firing 12-gauge (top) and 37/38mm (bottom) nonpyrotechnic rounds at windshield and side windows during training exercise. Penetration through front or rear vehicle windows is variable, depending on distance and angle.

Checking the results through the windshield and . . .

View from behind the windshield. As with the pyrotechnic projectiles, nonpyrotechnic projectiles can cause severe injury or death if they strike a human being.

penetrate a windshield out to 160 feet (48.8 meters) and a hollow-core door at 325 feet (99.2 meters).

Muzzle Blast Dispersion Rounds

Unlike the pyrotechnic and nonpyrotechnic rounds, this munition does not carry its payload in a separate projectile. Rather, the powdered chemical agent (or OC powder) is expelled directly out of the muzzle of the launcher.

These types of munitions are extremely well suited for close-quarters use, such as when protecting building entrances, gated entrances, or narrow approaches to entrances.

Though designed to be fired directly at crowds or rioters at close range—(within 30 feet or 9.1 meters)—these rounds should not be aimed directly at the face because of the danger to the eyes from flying particles. Rather, they should be aimed at leg level or lower. They may be used indoors or outdoors.

. . . side window.

No. T-21 muzzle-blast-dispersion round.

Muzzle-blast-dispersion round being fired through a 37mm gas launcher.

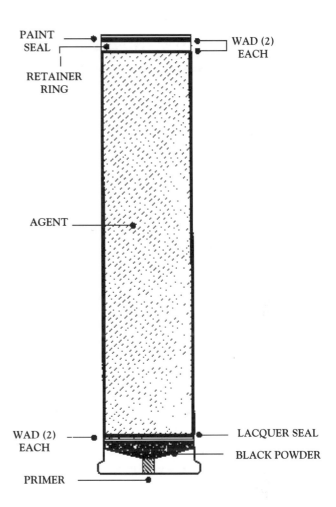

Illustration 9-3. Defense Technology T-21 muzzle-blast-dispersion round. This 37/38mm round expels powdered chemical agent or OC powder at close ranges. It can be used indoors or out. If possible, it should be deployed downward when using it in stairways to avoid the backflow of the agent toward the firer. (Courtesy of Defense Technology.)

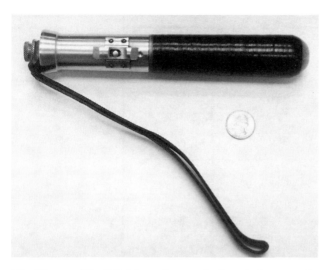

The 12-gauge "Gas Billy" fires what may be described as a muzzle-blast-dispersion round, albeit a small one with limited range. Shown here is an old Federal Laboratories' model. Similar devices such as the Brazilian AM-402 projector made by Condor SA-Industria Química are still being produced.

The Carbone Mark-1 experimental round combines elements of nonpyrotechnic projectiles and single-ball rounds.

NONPYROTECHNIC PROJECTILE MUNITIONS—MALFUNCTIONS

If a nonpyrotechnic projectile munition fails to fire after the shotgun or gas launcher's trigger has been depressed and contact has been made between the firing pin and primer, a second attempt should be made. If the munition fails to fire after the second attempt, then it should be carefully removed from the firearm and properly disposed of.

CARE AND MAINTENANCE

Nonpyrotechnic projectiles must be protected from high temperatures, humidity, and moisture. The manufacturers' recommended shelf life for individual munitions must also be taken into consideration.

CARBONE MARK-1

An interesting variation that combines aspects of both nonpyrotechnic chemical load projectiles and single-ball rounds such as Defense Technology's No. 23SB is the Carbone Mark-1 round. This experimental munition has reportedly been tested and evaluated for use by the U.S. Army.

PEPPER BAN™ PROJECTILES

Also of note is the increasingly popular Pepper Ban projectile systems from Jaycor Tactical Systems.

This system uses paintball-like, high-pressure, air-powered launchers to fire different types of frangible spheres that burst on impact, delivering a blast of OC powder or dye-marking liquid accurately out to distances of 30 feet.

Specialty-
Impact Munitions

Specialty-impact munitions are one of the solutions devised to deal with the problem of restoring the public peace when an out-of-control individual or crowd has disrupted it, *without* resorting to a direct application of deadly force to accomplish the mission.

The munitions illustrated and described in this section do not contain, nor are they classified as, chemical agent weapons.

This section is provided as an overview to aid personnel who may be tasked with procuring, testing, or employing munitions of this type. For more detailed information on specific products you should contact the manufacturers or distributors in your area or refer to *Specialty Police Munitions* by Tony L. Jones (available from Paladin Press).

A HISTORICAL PERSPECTIVE

As noted above, the new specialty impact munitions currently receiving so much attention are being marketed as less-lethal solutions to deal with the problem of restoring the public peace when an out-of-control individual or crowd has disrupted it.

A brief historical review indicates that the desire for this type of munition has not always existed. For instance, in

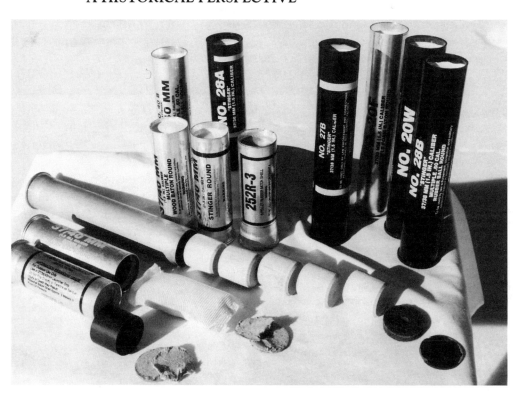

October 1795 approximately 80,000 Parisians stormed the Tuileries Palace, threatening to overthrow the newly formed moderate republican government. This government, known as the "Directory," was saved by a 26-year-old brigadier general who placed 40 cannons around the palace. This young general then ordered the cannons charged with extra powder and loaded with canister (grapeshot), nails, links of chain, and scrap metal. As the mobs came within several yards of the cannons, the general ordered the gunners to fire, killing or wounding uncounted numbers of attackers.

So it was that Napoleon Bonaparte saved the Directory by employing his own brand of "specialty impact munitions" to restore order.

It is interesting to note that even in the bloody era that produced the Reign of Terror, Napoleon feared being accused of excessive brutality for his actions. The social and political realities of the time, however, allowed his actions to be accepted as being necessary or, as he simply put it, "We were victorious, and all is forgotten."[1]

Fast-forward 194 years to Tiananmen Square in Beijing. After a series of student-led prodemocracy demonstrations spanning a seven-week period, the Chinese government declared martial law and ordered the protesters to end the demonstrations.

This order was ignored. On June 3–4, 1989, the People's Liberation Army, acting under the orders of Li Peng, brutally crushed prodemocracy supporters by employing machine guns and hardball ammunition. An estimated 3,000 to 5,000 people were killed, another 10,000 injured, and hundreds of students and workers arrested.

The protests were quelled and order was restored in a manner regarded by the communist leadership as both reasonable and necessary. In this case, however, being victorious did not mean all was forgotten. The times had changed, and the world had gotten smaller. The extremely violent suppression of the Tiananmen Square protest caused widespread international condemnation of the Chinese government that still resonates to this day.

As these two examples from history show, the ways in which a society chooses to deal with large-scale civil disobedience (or individuals whose actions, regardless of motivation, pose a risk to themselves and others) pretty much depend on the social, political, and economic climates and conditions prevalent at the time.

Currently, at least in the United States, the social and political climates of our modern, computer-driven, satellite-television-linked world seem to lean more toward tolerance for those disrupting the public peace and order than regard for those who are having their peace and order disrupted.

The riots that occurred in Los Angeles after the first Rodney King trial provide an excellent example of this bizarre "El Niño"-like social-climactic condition. As the world watched, rioters looted, pillaged, and burned while politicians from all levels provided running commentary on the "understandable reactions of the community."

Unfortunately, none of these pundits bothered to get the opinions of the members of the community who happened to own the vehicles, homes, and businesses that were being sacked, or who were themselves being brutalized.

Regardless of the causes, the existing social-political climate has definitely influenced many civilian governments to try to find more palatable and less lethal means of accomplishing the mission of restoring civil order—hence the development of modern specialty impact munitions.

WHEN TO USE THEM

There are many varieties of specialty impact munitions currently available that may be fired through 12-gauge shotguns and 37/38 and 40mm gas launchers.

While many of these specialty impact munitions do indeed provide viable options for use within the tactical environment when dealing with "lightly armed aggressive subjects," the policies that dictate when and how they will be employed must be decided well before they are used in actual situations.

Many police officers and administrators have been quick to adopt these devices, especially since they are often presented as a more humane, "less than lethal" alternative to standard penetrating-type munitions (bullets). The problem with this rush to adoption is found in the often-accompanying rush to confusion.

It is not uncommon for civilian police administrators to strongly (and openly) desire that the apparent "less than lethal" munitions be used whenever possible, and often, in their zeal to "soften" the perceived police response by reducing the likelihood of injury to an offender/suspect, they inadvertently place their own people at greater risk than is reasonable.

Having officers fire beanbag rounds at a suspect armed with a firearm is one example. When discussing this particular scenario, I have encountered more than a few officers who will argue that this tactic, if properly performed, can be used in a fairly tactically sound, efficient manner.

In most cases, the argument is made, highly trained and skilled officers, armed with standard weapons and ammunition, could be deployed safely, behind cover, with the suspect in their sights. The officer assigned to deploy the specialty impact munitions, also behind cover, can then attempt to disable or deter the armed offender by firing the specialty impact munition (in this scenario, beanbag rounds) at the offender.

Should immediate deadly force be required, the officers armed with standard ammunition can then apply it, stopping the threat and protecting the officer armed with beanbag ammunition.

Given the above conditions, this scenario sounds reasonable. Change any of the elements, such as the high skill level of the involved officers or the proper use of cover, however, and the resulting scenario becomes one in which the officers have placed themselves and the public at unnecessary risk.

While I firmly believe that flexibility, judgment, and discretion must be left to the individual officers dealing with the specific situation at hand, I am also acutely aware of the KISS principle,[2] Murphy's Law,[3] and Hicks Law.[4] And I have seen, first hand, too many instances proving that the best-laid plans of mice, men, and cops, do indeed often go awry.

My intent here is not to discount the potential value of these types of munitions. I am simply extremely weary of any equipment developed—and often adopted—in large measure to reduce liability exposure by lessening the chances of someone being able to claim "excessive force."

USE OF FORCE—REASONABLE AND NECESSARY?

To those who believe that police officers are required to perform their duty using the minimum amount of "reasonable *and* necessary" force, be advised that a more realistic standard to meet is one only of *reasonableness*. What this means is that a police officer doesn't have to pick the *best* way to accomplish the mission, only a reasonable way.

Reality demands the acceptance of this standard. Although our goal as police officers should always be to act in the best possible way in any situation, the truth is that when operating under the stress of a dangerous or violent encounter, this is not always feasible.

To determine if an officer (or officers) has acted reasonably in a specific situation, the following questions need to be answered:

- What did the officer do at the time?
- Was what the officer did at that moment in time reasonable?
- Would another officer, given a similar situation and circumstances, have acted in a similar manner?

If, after taking into account all of the circumstances, a determination can be made that the involved officer(s) acted in a reasonable way, then the standard has been met.

By throwing in the word *necessary*, we needlessly complicate the issue. Exactly what level of force was "necessary" in any situation can be argued endlessly based upon individual perceptions—and, as is too often the case, *mis*perceptions.

CONFUSION FACTORS

As noted above, the desire to lessen the liability profile of a department or agency is often one of the prime motivations for the adoption of specialty impact munitions.

Like most desires, however, this one is like a double-edged knife—it cuts two ways. The other liability-based consideration that must be considered is the potential for confusion on the part of a jury should an officer use a device classified as "less than lethal" and a death results. In such an occurrence, the opposing legal counsel will have enough ammunition of his own to cast doubt on the judgment, training, and competence of the officer, the department, or both.

The most important concern from our perspective, though, is the potential for confusion on the part of officers in the field who may be issued these devices.

Deciding to employ a less-lethal option during a lethal-force encounter is the gravest concern. Studies have indicated that police officers will often choose to place their own lives at greater risk than necessary while performing their duty because they desire so strongly to avoid having to injure another person—regardless of the circumstances.

Mistakenly loading the wrong munition into the shotgun while operating under the great stress generated by an ongoing conflict could also get the officer employing the device severely injured or killed should he fire a less-lethal device at a suspect presenting an immediate deadly threat and fail to immediately stop that threat.

Conversely, inadvertently loading and firing a slug or 00 buckshot round at a subject who presents less than an immediate deadly threat is also possible and carries its own weight of pain, penalties, and repercussions for the officer involved.

COMMITMENT AND TRAINING—THE CORE CONSIDERATIONS

In the final analysis, I believe that the most significant issue regarding the proper use and employment of specialty impact munitions is the same as with any other weapon or munition we are issued—training.

This means that any department or agency that procures any type of specialty impact munition *must* purchase enough to allow the officers who will be employing them to train with them properly before they are used in the field and to refresh their training periodically throughout the year. This requires a full commitment on the part of the agency and its members to the maintenance, training, and proper use of all equipment in the inventory.

Anything less is unacceptable.

A brief overview of some of the types of specialty-impact munitions specifically designed for civilian law enforcement use follows.

TYPES OF SPECIALTY-IMPACT MUNITIONS

Specialty-impact munitions generally operate on the same principles as the muzzle-blast-dispersion rounds. The difference is that instead of a payload of micropulverized chemical agent, these munitions disperse a rubber, wood, cloth, foam, or plastic projectile (or multiple projectiles) downrange. Normally these projectiles are neither configured nor propelled at velocities high enough to cause penetration of the body when used at recommended distances. Munitions are produced in 37/38 and 40mm calibers as well as for use in 12-gauge shotguns.

"MODI-PAC" 12-gauge rounds were one of the first commercially produced specialty-impact munitions. The small plastic pellets were skip-fired off flat surfaces in order to strike the legs of rioters. Firing directly at a person's body at closer ranges could result in penetration. In rare cases where this occurred, it was discovered that the plastic composition of the pellets made them extremely difficult to remove, especially since they didn't show up clearly on X-rays.

The manufacturers or distributors of the particular munitions you are interested in should provide you with all relevant data. They should also be able to provide your agency with training in the proper use of their product. Barring that, they should be able to put you in touch with an authorized agency, company, or legitimate trainer that can certify you or your personnel in its use. This is necessary because, although many of the munitions are simple enough to use if you take the time to read the instructions and understand how to make a shotgun or gas launcher go "bang," you must do your best to ensure that there are no unpleasant surprises waiting to be discovered when using the rounds in the field. Receiving documented training and being certified as a user or instructor for the particular munitions you'll be using also help to establish the oh-so-important paper "back trail" that will provide

more layers of liability insulation for you and your department should a problem develop.

Rubber Projectiles

Rubber projectiles have been produced in many shapes, weights, and designs. The most common shapes for projectiles used in specialty impact munitions are ball, cylindrical (baton), and a fin-stabilized hybrid of both.

Munitions using rubber-ball rounds may employ single or multiple projectiles. These rubber balls may vary in diameter from .32 to .60 caliber. The smaller caliber munitions are usually loaded with more individual projectiles.

Effective ranges and velocities of these types of munitions vary.

Single Rubber Ball Rounds

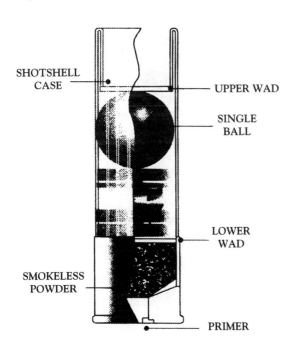

Illustration 10-1. No. 23SB .69-caliber single-ball round. This round travels at an initial velocity of approximately 900 fps and has a maximum effective range of about 70 feet (21.4 meters). Single-ball rounds such as these may be skip-fired toward threatening individuals or crowds. Direct-firing these rounds at an individual or into a crowd may result in serious injury or death! These particular rounds, such as the 12-gauge baton rounds, are also very well suited for "porting" (breaking) windows. (Courtesy of Defense Technology.)

Multiple Rubber Ball Rounds

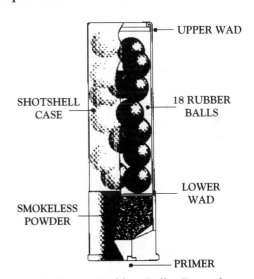

12-Gauge Rubber Pellet Rounds

Illustration 10-2. Such rounds as the No. 23RP and No. 23HV hold approximately eighteen .32 caliber rubber balls. The primary difference between these two particular rounds is found in the velocity of the projectiles as they exit the barrel and head downrange. The No. 23HV (high-velocity) round, loaded with approximately .18 grams more smokeless powder propellant than the No. 23RP, pushes its rubber ball payload out of the business end of the barrel at about 900 fps. This makes the round a bit more effective than the No. 23RP, which produces 400 fps, when you "reach out and touch someone." Warning: Either munition may cause serious injury or death if fired at extremely close range! (Courtesy of Defense Technology.)

37/38/40mm Rubber Ball Rounds

Left: The 5 1/2-inch (14-centimeter) long No. 27A Stinger round holds approximately one hundred and seventy-five .32-caliber rubber balls, while the No. 28A holds approximately twenty-seven .60-caliber rubber balls. Larger versions of these particular munitions (No. 27B and No. 28B Stinger rounds) that hold more projectiles of the same calibers and offer slightly higher velocity are also available from Defense Technology. A third variation of these multiple rubber ball projectile munitions (No. 40A and 40B) is also available for use with 37/38mm gas guns as well 40mm launchers. All of these rounds (No. 27A/B, No. 28A/B, No. 40A/B) can be fired directly at violent individuals or crowds, or skip-fired off a flat surface, such as the ground or a wall. The larger caliber rounds tend to be more effective against subjects dressed in heavy or layered clothing. Warning: Any of these munitions may cause serious injury or death if fired at extremely close range!

12-Gauge Rubber Fin-Stabilized Rounds

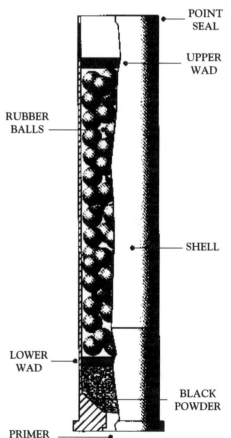

Illustration 10-3. No. 27A Stinger round. (Courtesy of Defense Technology.)

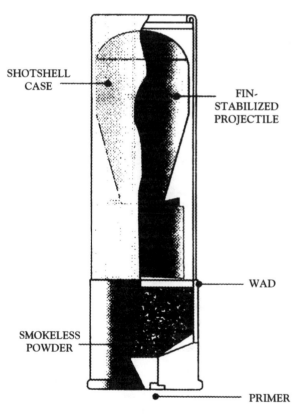

Illustration 10-4. Single-projectile, fin-stabilized rubber rounds are also available. Shown in this illustration is the No. 23FS rubber fin-stabilized round produced by Defense Technology. This round is designed to be direct fired at threatening individuals or crowds. The projectile, approximately 1.7 inches (4.3 cm) long, travels at an initial velocity of 500 feet per second (fps). Warning: This projectile may cause serious injury or death if fired at close range!

37/38 and 40 mm Multiple Foam Rubber Baton Rounds

Multiple foam-rubber baton munitions, like the rubber ball rounds, are designed to be fired directly at violent individuals or crowds. The 8-inch (20.3-centimeter) long No. 20F 37/38mm round shown above on right fires five foam-rubber projectiles at about 320 fps, out to an effective range of approximately 50 feet (15.2 meters), though projectiles can travel up to 100 feet (30.5 meters). The 4.8-inch (12.2-centimeter) long No. 40F 37/40 mm round on the left fires three foam-rubber projectiles at approximately the same velocity and distances and can also be fired through 40mm launchers. Both rounds launch projectiles similar in length, diameter, and weight.

Note the differences between the foam-rubber baton round (left) and the solid-rubber baton round. The foam-rubber baton rounds are designed to "give" more upon impact, lessening the likelihood of serious injury.

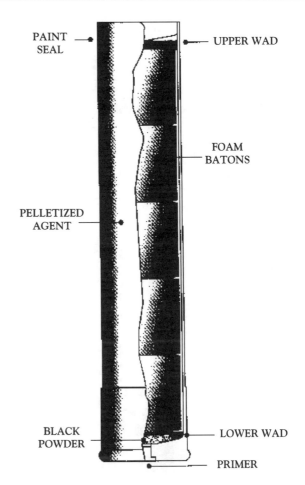

Illustration 10-5. No. 20F Foam-Rubber Multiple Baton Round. Each baton is 1.5 inches (3.8 centimeters) long and 1.5 inches in diameter, and weighs 0.60 ounce (17 grams). (Courtesy of Defense Technology.)

Wood Projectiles

Wood projectiles are usually produced in cylindrical shapes of various lengths and referred to as "baton" rounds. Munitions using wood baton rounds may employ single or multiple projectiles. These wood batons may vary in diameter from .62 caliber (1.6 cm) to 1.35 inches (3.4 cm). Lengths of individual projectiles tend to differ, depending upon overall size of munition and number of individual projectiles used.

Range and velocity of these types of munitions vary, though their effective range usually falls within 50 to 60 yards (45.7 to 55 meters). These munitions are intended to be used as skip-fire impact projectiles under all but deadly force circumstances.

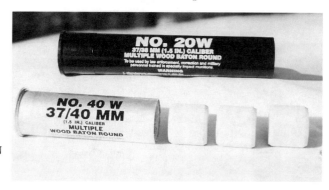

12-Gauge Single Wood Baton Rounds

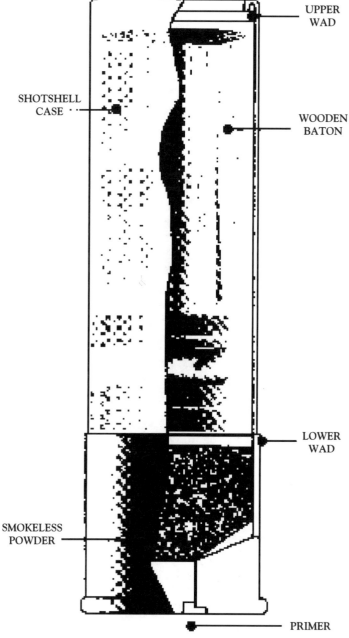

Illustration 10-6. The No. 23WB Single Wood Baton Round. Rounds like this 12-gauge projectile are designed to be fired through a standard 12-gauge shotgun. It may be skip-fired into threatening crowds or toward a threatening individual. Under extreme conditions it may be fired directly at a crowd or subject, though it must be understood that use in this way constitutes deadly force. This round is also very well suited for porting windows during a barricade or hostage situation. (Courtesy of Defense Technology.)

37/38 and 40mm Wood Multiple Baton Rounds

Multiple wood-baton rounds are not recommended to be fired directly at violent individuals or crowds. The 8-inch (20.3-centimeter) long No. 20W 37/38mm round fires five wood projectiles at approximately 280 fps out to an effective range of approximately 50 yards (45.7 meters) when skip-fired, though projectiles can travel up to 100 yards (91.4 meters). The 4.8-inch (12.2-centimeter) long No. 40W round fires three wood projectiles at approximately the same velocity and distances and can also be fired through 40mm launchers. Both rounds launch projectiles similar in length, diameter, and weight. Unlike the foam rubber rounds, the wood batons should only be skip-fired into crowds or toward individuals in all but deadly force situations. Warning: Firing a wood baton round directly at a subject may result in serious injury or death!

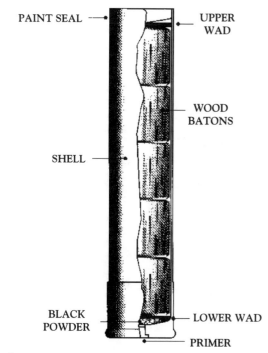

Illustration 10-7. No. 20W Wood Multiple Baton Round. Each baton is 1.35 inches (3.4 centimeters) long, 1.35 (3.4 centimeters) inches in diameter, and weighs 0.77 ounce (22 grams). (Courtesy of Defense Technology.)

Beanbag Rounds

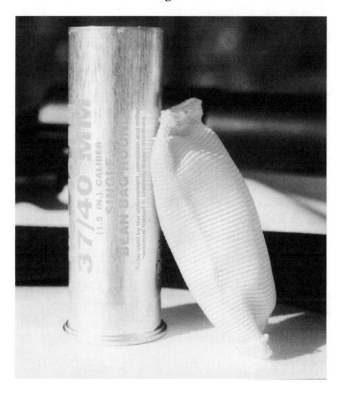

Beanbag rounds are produced in 37/38 and 40mm calibers for use in gas guns and launchers, and in 12 gauge for use in standard 12-gauge shotguns.

The projectile itself is exactly what its name implies—a small cloth bag filled with pellets, beans, or sand. These bags may be square or tubular ("sock shaped"). The sock-shaped munitions are currently gaining popularity because their trajectory tends to be more consistent and predictable.

One of these rounds, the new 23DS 12-gauge beanbag round from Defense Technology, is purported to be one of the most accurate rounds of this type. Produced in an aerodynamic, tear-shaped design with a multiple-tail drag-stabilization (DS) system, the 23DS is reportedly capable of achieving accurate hits out to distances of 60 feet (18.3 meters) and beyond.

The square-shaped munitions, on the other hand, open while in flight before impact, delivering their energy to the objective without penetrating it. There have reportedly been instances where the square-shaped rounds have impacted on edge, however, causing serious injury and death.

Beanbag rounds are designed to be direct fired—to the body, not the head—at violent or aggressive individuals.

12-Gauge Beanbag Round

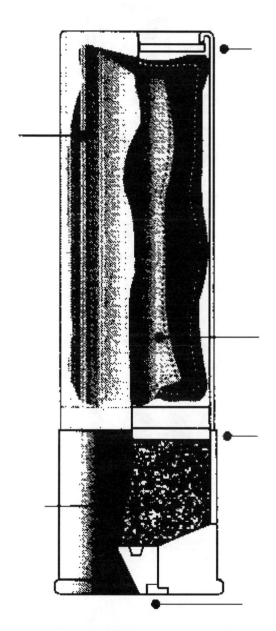

Illustration 10-8. No. 23BR Beanbag Round. This round fires a single beanbag at 300 fps out to 150 feet (45.8 meters), with an effective range of approximately 50 feet (15.2 meters). This projectile is made of white cotton canvas approximately 2 x 2 inches square and filled with #9 shot. It weighs approximately 1.44 ounces (41.0 grams). (Courtesy of Defense Technology.)

37/38 and 40mm Beanbag Round

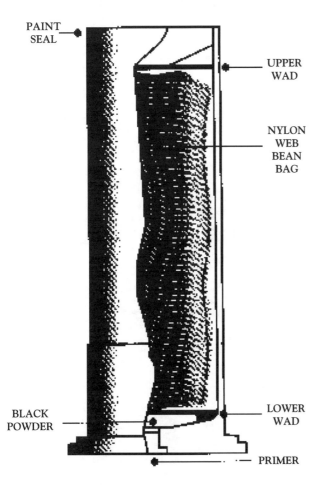

PAINT SEAL

UPPER WAD

NYLON WEB BEAN BAG

BLACK POWDER

LOWER WAD

PRIMER

Illustration 10-9. No. 40BR Beanbag Round. This round fires a single beanbag at 220 fps out to 100 feet (30.5 meters), with an effective range of approximately 50 feet (15.2 meters). The projectile is made of tubular webbing and filled with silica sand. It weighs approximately 3.5 ounces (100 grams). (Courtesy of Defense Technology.)

Other Specialty-Impact Munitions

Because of the continued and growing interest in specialty impact munitions in both the civilian law enforcement and military industries, new products are constantly being developed, tested, and refined. Variations of these products are also being worked on throughout the world, and future developments in this field are sure to provide many surprises and,

perhaps, some solutions to the problems that are fueling the interest.

In addition to the various single and multiple rubber projectiles, beanbag rounds, and batons made from wood, rubber, and foam munitions shown in this chapter, there is a multitude of new devices offering fresh approaches as well as some twists on a few older concepts. One such device is the 12 x 12-foot square Netgun from Capture Systems, Inc. This device can be fired at subjects up to 45 feet away and ensnare them in a 200-pound tensile-strength nylon net. Another approach along similar lines is the 12-gauge Bolo Capture Projectile that propels three 46-grain rubber projectiles connected by high-strength cords out to effective distances of 20–40 yards (18.3 to 36.6 meters). Should one of these balls strike the intended recipient's legs, inertia causes the other balls to wrap the cord around the fleeing suspect's legs and trip him up. Interesting concept, but the potential for Mr. Murphy Mayhem seems a bit high. . . .

Dye-staining rounds; CO_2; N_2 and compressed-air launchers; sponge[5]; soft, rubber-like, ring-shaped kinetic energy projectiles; and as many variations as you can care to imagine are out there to choose from, as well as a number of other options designed to stop less than desirable behavior through less-lethal means.

SPECIALTY-IMPACT MUNITIONS — MALFUNCTIONS

Should a specialty impact munition fail to fire in training after the shotgun/gas launcher's trigger has been depressed and contact has been made between the firing pin and primer, a second attempt should be made. If the munition fails to fire after the second attempt, then it should be carefully removed from the launcher and properly disposed of.

Should the munition fail during an operation, then another round should be immediately loaded and fired, unless the threat to the officer has escalated to one of deadly force. In that case, the officer should immediately switch to another weapon.

CARE AND MAINTENANCE

Specialty-impact rounds, like any other type of munitions, must be properly stored and maintained. Care should be taken when handling them to avoid dropping or damaging them. They should be kept in the packaging provided by the manufacturer and stored in a secure, moisture-free environment. Expiration dates must be noted and stocks rotated as needed.

NOTES

1. Owen Connelly, *Blundering to Glory: Napoleon's Military Campaigns* (Wilmington, Del.: Scholarly Resources Inc., 1987).
2. Keep it simple, stupid.
3. Whatever can go wrong will go wrong, at exactly the worst possible moment.
4. According to Hick's Law, from a reaction-time standpoint, the human being will perform in the fastest manner when there is only one option to choose from. If you add a second option, you increase this time by 58 percent.
5. Defense Technology has just introduced a 40mm spin-stabilized sponge round called eXact impact.

Aerosol Subject Restraint Sprays

The term *aerosol subject restraint* (ASR), reportedly first coined by William J. "Doc" McCarthy, is used by the chemical defense spray industry to describe products using OC.

In the foreword to Doug Lamb's book, *Pepper Sprays*,[1] Doc McCarthy explains that the designation ASR was inspired by the need to differentiate between the "natural" OC agent and the chemically manufactured "tear gas"[2] agents CN and CS, which are also used in aerosol form.

It should be noted, however, that some chemical aerosol sprays currently being marketed contain a combination of agents, the most common being a mixture of OC and CS, or OC and CN.

For the purposes of this book and to avoid confusion in general, these combined sprays will also be referred to as ASRs.

Inset: Aerosol subject restraint sprays sometimes use a combination of OC and CN or OC and CS. The Freeze +P units shown here use a formulation of CS/OC. They are very effective.

SPRAY PATTERNS

ASRs use three primary spray patterns determined by the design, size, and shape of the nozzle of the unit. When the button or actuator is depressed, the pressurized agent within the canister is forced out through the nozzle in either a conical mist, ballistic stream, or fog pattern.

Left to right: Fogger, conical mist, and ballistic stream-type nozzles. The unit shown at the far right dispenses its "pepper foam" payload by way of a ballistic stream pattern actuator. The author prefers a ballistic stream pattern for most police field applications.

Conical Mist

A conical mist spray is very effective at average distances. Depending on the size of the unit being used, these distances generally range from 4 to 10 feet (1.2 to 3 meters), with some larger units being effective out to as much as 15 feet (4.5 meters). OC in particular works best in mist form.

When applied correctly (i.e., in short bursts), the mist spray will produce faster and more dramatic effects than the ballistic stream. This is because the agent is delivered in a purer form to the sprayee and does not require as much time to "drop" through the carrier and take effect. In its atomized state, the agent is also more easily inhaled. This allows it to affect the airway and bronchial passages far more efficiently than when delivered in the ballistic stream.

The downside, however, is that the effective range of the spray is decreased, while the potential for collateral damage is greatly increased. It is very easy to contaminate others—including yourself, fellow officers, and innocent bystanders—when using this spray pattern, especially in windy conditions.

When using it indoors, the effect may be magnified even more, as the atomized agent tends to "wander." This should be kept in mind if you are operating in a heavily congested environment or building. I know of several instances when certain areas in a police station or barracks were cleared of

Conical mist spray pattern.

Ballistic stream spray pattern.

all personnel by a mysterious application of atomized OC. I have not done a study on the matter, but in my experience a rash of these unprofessional incidents tend to happen right after initial issuance of ASRs to departmental members—especially on slow nights.

Although occasionally providing a modicum of perverse entertainment, it must be remembered that the use of these products in this manner may have unanticipated, possibly even tragic, consequences. It is also an egregious violation of the safety rules for handling less-lethal weapons (see Chapter 1).

Ballistic Stream

The ballistic stream pattern provides for an accurate, controlled application of the agent to the face and eyes of the sprayee. Formulas combining OC and CS tend to work very well when delivered in this manner.

Besides improved accuracy, the stream also allows you to reach out and touch someone at ranges far greater than those attainable with the conical spray pattern, generally from 12 to 15 feet (3.6 to 4.5 meters) with an average-sized unit.

The ballistic stream also tends to produce less collateral damage than the conical spray, making it a more attractive choice given the dynamic nature of police work and the heavily populated areas many of us operate within.

The downside here involves the slightly longer time it takes for the agent to drop through the necessarily heavier carrier, allowing the effects to begin. Although the eyes will generally slam shut instantly, the other debilitating effects may not kick in for as long as 5–10 seconds. That is a lot of time should you become engaged in a wrestling match with a screwball.

If these considerations are addressed in training, however, and if the ASR is used in a tactically efficient manner, I have found that the delay is an acceptable trade-off for the improved accuracy, distance, and control afforded by the stream. I reached this opinion over a two-year period while field-testing a variety of agents and spray patterns in a highly active environment.[3] After numerous real-world applications, as well as several direct personal

MSI's MK-IV pepper foam uses a 10-percent OC formula. While appearing significantly different from resin-based formulations, the foam is dispensed by way of a ballistic stream pattern actuator. Extremely effective both physically and psychologically, though the thick foam does not consistently induce respiratory effects because of the reduced potential for inhalation. There is a potential, however, for the sprayee to scoop some of the foam from his own primary objective area (POA) and attempt to deliver it to the sprayer's POA.

Smaller ASR equipped with fogging pattern-type nozzle.

exposures under controlled circumstances, I stand by this recommendation.

Fogging pattern

Fogging spray patterns resemble the output of aerosol-type fire extinguishers. The fine, fog-like mist that is produced is very efficient, and may be instantly effective at ranges up to 20 feet (6 meters), depending on the size and type of the unit being used.

Extremely useful for employment against

multiple subjects. The fogging spray pattern, in addition to clearing people from an area, can also be used to temporarily deny an area to "unfriendlies" by way of the chemical barrier that it produces.

The same problems attributed with the use of conical spray patterns (agent wander and collateral damage) also exist with the fogging pattern.

An additional consideration that must be weighed when using fogging-pattern actuators is the decreased number of shots available per unit. This occurs because the fogging pattern puts out a large amount of agent under high pressure with each burst, quickly depleting the payload.

CARRIER AGENTS

Carrier agents are the means by which the actual chemical agents are delivered to the objective. For example, oleoresin, literally an "oily resin," is generally dissolved in a carrier such as alcohol, ketone, Dymel-22,[4] or water.

Nonflammable carriers (such as water or mineral oil) are currently the preferred "vehicle," especially since the well-publicized case involving the New York Police Department's Emergency Services Unit (NYPD ESU) occurred several years ago.

In this case, a subject resisting arrest was sprayed with an agent using an isopropyl alcohol carrier. When this didn't produce the immediate desired effect, officers used a TASER.[5] An isopropyl alcohol-contaminated shower curtain then ignited and the suspect was burned. It is believed that the ignition occurred as a result of either the initial TASER firing or subsequent arcing once the contacts were in place.

BELT-CARRIED UNITS: PRESENTATION, STANCE, AND GRIP

The presentation of the belt-carried ASR unit should be kept as simple as possible. Many departments currently advocate that the holster/ASR unit be placed on the duty belt directly in front of the service pistol/holster. I have found that this positioning of the unit makes both training and actual accessing of the ASR faster and simpler because of the familiarity of the movements when the ASR is employed with the dominant hand.

How the Agent Is Delivered

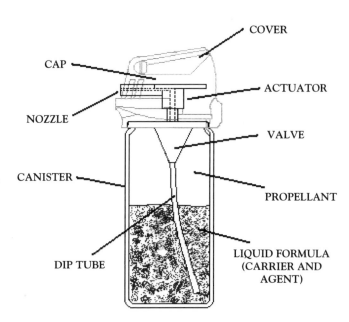

Illustration 11-1. The agent is suspended in a liquid-based carrier vehicle. The liquid is under pressure inside the canister. When the actuator is pressed, the liquid is forced through the tube and out the nozzle.

NOTE: The holster/ASR unit must not interfere with a clean presentation of the pistol!

Some departments advocate using the support or nondominant hand to access and administer the ASR spray. Common arguments made for using the nondominant hand are as follows:

1. The ASR can be extended fully toward the sprayee's face with the nondominant hand while the dominant hand remains free to control the handgun or deliver a strong blow if necessary.
2. By keeping the nondominant side of the body forward, the familiar "interview position" stance is maintained, allowing the officer more mobility to spray and avoid the suspect.
3. Extending the dominant arm for spraying forces the officer to leave the pistol exposed and unprotected.

I prefer to hold the canister in the dominant hand because it feels more natural and better accuracy is possible.

Regardless of where you position the holster/ASR unit or which hand you use to present and employ it, you must be able to access the canister quickly and without interference. Holster flaps or strap devices should be easily opened, and the unit should be removed quickly and brought into full presentation position cleanly, while employing a solid grip on the device.

Your stance, always a critical component when you are tactically engaged, should be wide enough to provide stability yet still allow you to move quickly in any direction.

Tactically reholstering the unit must also be taken into consideration and practiced. As with the pistol, the reholstering should be done with one hand, while keeping your eyes off the gear and on the suspect.

Additional consideration must be given to the practice of switching from the ASR to a handgun, should the situation suddenly escalate to the level where deadly force is justified.

Specific ASR presentation, reholstering, and transition drills are discussed in Section II. Three dominant-hand presentation stances are demonstrated by Paul Damery in the accompanying photographs. To simplify matters, I am going to refer to these as positions 1, 2, and 3.

Position 1

The ASR, grasped in the dominant hand, is held back toward the sprayer's body. The support (nonspraying) arm is forward in a defensive posture to ward off blows and to keep the intended sprayee from grabbing the unit from you. The stance is wide and balanced. This technique, perhaps best suited for close-in, defensive-type work, is the least tactically desired position when using an ASR offensively because it violates the time/distance/violence principle of close-quarter engagements by encouraging the operator to get too close to the sprayee. If used at greater distances, this position adversely affects accuracy because of the awkward stance and poor placement of the controlling primary hand.

Position 2

The ASR, grasped in the primary hand, is held extended toward the sprayee. The stance is, again, wide and balanced. This position, which feels very natural under stressful conditions, allows the ASR to be aimed much like the pistol is when employing the point shooting technique.[6] In this photograph, the support hand is held back toward the sprayer's body in a defensive ready position. The support hand may also be held out to the side for balance.

Position 3

Actuator Finger Position

There are two schools of thought when it comes to deciding which digit— the index finger or thumb—is best for pressing the top-mounted actuator. Deciding which is best for you will require some experimentation: the grip on the canister is different depending on the digit used. Once you find the method that suits you best, you should use it consistently.

Index Finger

This index or "trigger finger" is in position to press the actuator. The use of this finger is both comfortable and familiar, most people having had some experience performing the same action with common household cleaning products, spray paint canisters, or some other product. The grip must be secure.

Thumb

A good argument can be made for using the thumb to trigger the top-actuator-mounted ASR. This hold is often easier to establish when presenting the ASR from the holster and generally provides a more secure grip on the canister. Canisters with a flip-top safety cover (as shown) are recommended to help prevent unintentional discharges.

The canister, grasped in the primary hand, is held extended toward the sprayee. The index finger is placed on the top-mounted actuator. The support hand is placed over the back of the primary hand for stability and support. The stance is wide and stable. Many firearms-trained operators find this technique for presenting and holding the ASR familiar because it is similar to support positions used when employing the sidearm. Some officers feel that it also appears more professional than the other holds, and signals a greater degree of seriousness of intent on the part of the officer to the sprayee as well as other onlookers. The likelihood of an officer assuming this position under stressful field conditions depends totally on the amount and quality of training time he has put in.

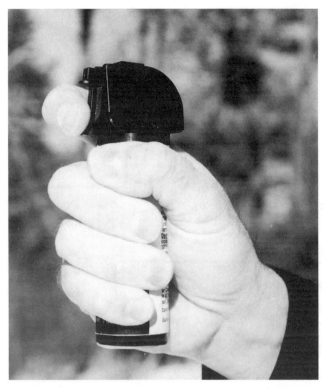

When using a front-trigger-mounted actuator, such as the one shown on this Winchester unit, the normal trigger finger is used. This type of actuator provides for a commonality of training for officers because its design is similar to that found on pistols.

This officer is using a fogger-type unit with a one-hand hold. The larger fogger units are intended for use at greater ranges and are extremely well suited for use against multiple subjects. The unit should be held extended toward the objective area while administering application of agent.

Presentation, Stance, and Grip—Summary

I have used all of the techniques demonstrated in this chapter under actual field conditions. Each is efficient and, when used properly, viable. Regardless of which technique is used, you must remember that as police officers we will be employing the agent in an *offensive* rather than defensive manner.

The ASR, like any weapon we employ, is most often going to be used to assist us in *controlling out-of-control behavior* and *taking someone into custody*—not, as a citizen might employ it, strictly for defensive purposes.

This fact will dictate that—for our purposes—the agent should be used aggressively, preferably using the element of surprise when possible and always from an advantageous position.

HOW TO ADMINISTER IT

The agent should be administered in short, quick bursts, each lasting anywhere from 1/2 second to 2 seconds. The subject's eyes, nose, and mouth are the primary areas to aim for. If possible, you should shake the canister immediately before spraying. The canister must be held upright to ensure proper operation.

Different techniques may be used to ensure adequate coverage of the objective area. Several short bursts may be applied, using minimal movement of the canister in the primary hand to produce various sweeping spray patterns, such as those illustrated here.

The practice of "over spraying" a subject using any of these techniques must be avoided. If you spray a subject with an ASR and hose him down with the carrier-based agent, what you are doing in effect is washing the agent away. This will then increase the time required for the agent to drop through the carrier and take effect.

So while it may be psychologically reassuring to apply a little more agent than is necessary—especially to a hostile, aggressive suspect—*less* is actually *more* effective.

Verbalization

Clear, direct, and loud verbal communications should be used by the officer, if feasible, when he is

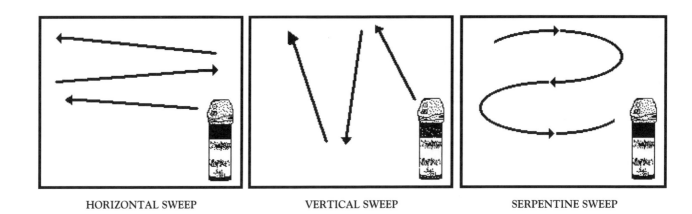

HORIZONTAL SWEEP   VERTICAL SWEEP   SERPENTINE SWEEP

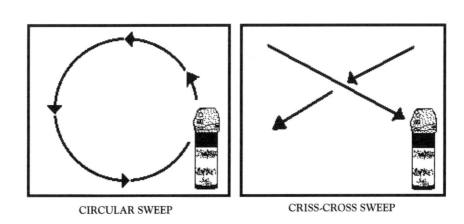

CIRCULAR SWEEP   CRISS-CROSS SWEEP

Illustration 11-2. Horizontal sweep, vertical sweep, serpentine sweep, circular sweep, criss-cross sweep. The circular sweep is considered by many to be the most efficient method to apply the agent to a moving subject.

involved in a critical use-of-force situation. This is done not only to get the offender to submit and comply but also to alert other officers of the situation. Witnesses will often recall whether any verbal commands or warnings were given, and, depending on the situation, this could make quite a difference should you eventually end up in a courtroom.

In regard to the employment of ASR sprays, some phrases that may be used include (but are not limited to) the following:

- "Police! Don't move!"
- "Get back—don't make me spray you!"
- "Stop!" or "Back off!"
- "Don't rub your eyes!"
- "On the ground before you fall."
- "Hands behind your back!"

- "Breathe slowly."
- "I will help you if you stop fighting."

Officers should also be trained to alert other officers before administering an application of OC, especially if several officers are in the process of trying to subdue a violent, resistive suspect. A clear and forceful shout of "OC!" is usually sufficient—and greatly appreciated by all involved officers.

Static Applications

While performing your duty, there may be times when you will need to apply an ASR spray to an offender under static, or nonmoving conditions.

A suspect who has been advised that he is under arrest and has resisted in some physical manner, who

Applying an ASR to a person under static conditions.

may have also made verbal threats or physically threatening gestures, may indeed warrant an application of OC or OC/CN/CS combination.

The justification for the use of a chemical agent at this level of force will depend on your training and perceptions, as well as on your department's use-of-force policy.

When employing ASR sprays under these conditions, it is recommended that the subject first be *asked* to comply with your directives and submit to the arrest. If he refuses, the subject should then be *advised* to comply and warned that he will be sprayed with the agent should he continue to resist. The final stage is to *order* the subject to submit. If the subject submits and then follows your verbal commands, you should have him assume some type of control position before handcuffing (e.g., kneeling, ankles crossed, hands on top of head; or prone on the ground, feet spread wide, hands extended to the sides, palms up). Should the subject continue to resist in a physically or verbally threatening manner, a 1/2-to-1-second burst of spray should be applied to the facial area.

In my experience, static situations are best handled in this manner. It is important that you say what you mean during the incident, as well as mean what you say. By this I mean that you should not continue to advise the subject or resort to threats to try to get him to comply once you have reached the final stage of "order." If he continues to resist after *order*, you spray.

If you keep yourself and the situation under control in this way, you shouldn't have any problems justifying your actions either to supervisors or in written reports. A calm, reasonable, matter-of-fact manner will also leave witnesses impressed—or at least with nothing much to complain about. (This is assuming that no family members or accomplices of the subject are present; if one or more are present, their presence and potential dangers must be considered when choosing to use any level of force.)

Once the subject has been sprayed, you should immediately advise him that you will help him through the experience if he does what you tell him. This is done from a distance while you wait for the agent to take effect.

When the agent takes effect, the suspect should be directed to assume a control position. Due to the disorienting effects induced by the agents, prone control positions may be preferable to other options. Any commands given should be loud, clear, and simple. The commands should also be repeated until the suspect complies; as the pain increases the suspect's ability to focus on your words may be hampered. If he continues to resist, then open-hand control tactics or other appropriate levels of force may be applied. Always approach subjects from the side or rear as opposed to the front (or "inside") position when preparing to take a subject into physical custody.

The suspect must be taken into physical custody, handcuffed, searched, and transported back to the station before you begin to assist him with decontamination. You must remember that once sprayed, the now disabled suspect's safety is very much your responsibility.

Some departments use what is often referred to as an "OC administrative warning." This printed form is read by the arresting officer to the person exposed to the agent. The subject is told what type of agent he was sprayed with, how long the effects should last, and that he will be assisted if he cooperates with the arresting officer. The arrestee is further advised that the effects of the agent may mask other medical conditions including overdoses or toxic levels of such drugs as cocaine, amphetamines, barbiturates, PCP, opiates, or alcohol. The subject is then asked if he is under the influence of any of these substances, or if he has any other preexisting medical conditions or allergies. This information should be recorded and taken into consideration by the arresting officer while processing the arrestee.

Of course, regardless of the answers to these questions, the subject should be carefully watched for any signs of unusual or extreme adverse reactions while in your custody. If any severe symptoms do occur, you must provide medical assistance to the subject as soon as possible.

Remember when writing your report to detail the facts that led to the application, as well as any escalation or de-escalation of force, decontamination efforts, and medical aid rendered.

NOTE: Always maintain your awareness of the potential threat the suspect presents to you. Do not bet your safety on the effectiveness of any particular ASR, other intermediate weapon, or physical control tactic. This is especially true when dealing with individuals who are under the influence of alcohol, drugs, or certain mental disabilities that do not allow them to recognize pain. Some subjects will also have a natural resistance to OC, causing the manufacturers to acknowledge that most ASR sprays are effective only 72 to 83 percent of the time. On several occasions I have seen handcuffed suspects suffering from the effects of a direct exposure of an ASR spray continue to fight. In one of these instances, the suspect spit a mouthful of OC/CS spray into the face and eyes of one of the officers trying to subdue him.

Dynamic Applications

Applying an ASR to a person under dynamic conditions.

Always use care when handling a subject who has been sprayed with an ASR. If the situation allows, the agent should be allowed to dry before transporting a contaminated subject in your cruiser to prevent collateral contamination. This should only take a few minutes in most cases. When transporting a contaminated subject, it is a good idea to open the cruiser's windows to allow for cross ventilation. Rubber gloves, if available, should always be used when handling any suspect. Keeping a pair in the bottom of your handcuff case, as shown above, is a great idea. They will be immediately available as soon as the handcuffs are removed. Keeping a pair in an empty 35mm film canister in the glove box, briefcase, or jacket pocket is another good idea that many officers use.

The term *dynamic applications* refers to those times when you decide to use an ASR spray against a subject while he is moving, you are moving, or you both are moving. This type of situation is commonly encountered on the street and must be prepared for through realistic, continuing training. In addition to developing the skill needed to be able to apply the agent accurately to a moving subject's primary objective area (i.e., face), several other considerations must be trained for. The most serious involves training the officer to respond appropriately to an escalated threat after the ASR has been drawn. Unfortunately, many departments never prepare their people for this type of scenario, and the consequences can be tragic.

Training to correct this problem is relatively simple and can be accomplished during departmental firearms training periods. A specific transition training drill is outlined in Section II.

Another consideration that must be prepared for while operating in an "open" or dynamic environment is collateral damage to personnel other than the intended sprayee.

Having seen this occur on both small and large scales, I can attest that it can be problematic. I was present when a crowd of uninvolved spectators at a high school sporting event caught some "overspray" directed at a few overzealous fans fighting in the stands. More than once I have seen a police officer or officers getting dosed by a co-worker while wrestling a suspect to the ground because the sprayer didn't sound off with a warning before letting loose. On another occasion I witnessed a mall full of people coughing and tearing because of conical "spray drift" after a physically resistive suspect was taken into custody. Do these things happen? Of course they do. Is it objectively unreasonable when it occurs? Well, yes and no.

When it happens, the innocent people who suffer the effects will generally feel it is unreasonable. You must remember that very often these people have no idea of exactly what is happening to them. Nor, in most cases, will they understand that the agent's affects will wear off fairly quickly. Elderly people or very young children may also be exposed in instances such as these, and the effects, psychological and physiological, can be harsh. Panic may result.

In my opinion, if innocent people are unnecessarily exposed to the agents because of improper application techniques, then, yes, the actions of the person employing the ASR may qualify as unreasonable.

If, however, the officer uses reasonable judgment and employs the ASR properly and professionally—as he has been trained to—taking as much care as possible to ensure that only the intended sprayee is affected, and someone other than the intended

sprayee is inadvertently exposed during a dynamic encounter, then, no, it is not unreasonable. The officer's clearly written report documenting the incident and his actions will show why.

The primary differences, then, between a reasonable action and an unreasonable action can be pared down to three components: (1) training, (2) reasonable judgment, and (3) report writing skills.

Having the knowledge and ability to decontaminate uninvolved personnel quickly will also weigh in on one side or the other if the incident results in litigation. Simply calling an ambulance to respond to the scene with saline can greatly improve the appearance of professionalism and lower the liability exposure of the officer and department, as opposed to advising the affected personnel to "suck it up" and "move out smartly with a purpose." Not that I've ever seen that done. . . .

Primary Objective Areas

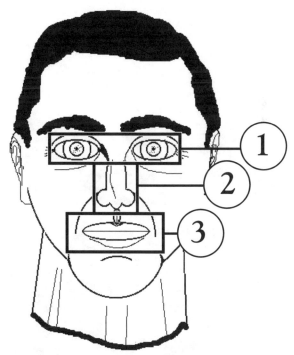

Illustration 11-3. The primary objective areas when using an ASR, in order: the eyes, nose, and mouth. When using any ASR, you must be careful not to spray directly into the eye at close distances (less than 3 feet [9 meters]). This is especially important when using an ASR with a ballistic stream because a "hydraulic needle" effect may cause permanent structural damage to the tissues of the eye.

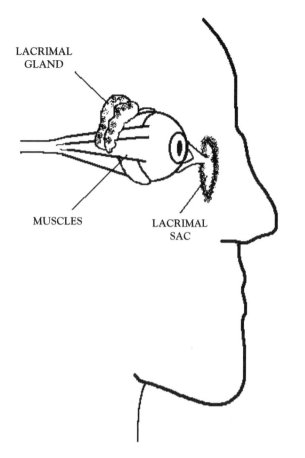

LACRIMAL
GLAND

MUSCLES

LACRIMAL
SAC

Illustration 11-4. The eye.

Subjects who are sprayed may not simply cease their attack and fold up into a "ball of silence." Though the initial effects of the agent may cause the eyes to slam shut, the more painful effects of the agent may take a few seconds to "drop" through the carrier. One of the benefits of experiencing the effects of the spray personally is realizing that instant incapacitation is not necessarily going to result.

WHAT YOU CAN EXPECT

When OC was first introduced in its present formulation, it was generally heralded as a panacea for practically all of the use-of-force problems faced by the law enforcement industry. As I indicated before, some advertisements for the product went so far as to suggest that it was as effective as the firearm for stopping a charging assailant armed with a knife.

After many years of actual use, the results are still being tabulated. However, we now have a more realistic perception of what ASRs will and will not do.

Some manufacturer's claims aside, *what you should not expect when applying an ASR spray to a subject is instant and total incapacitation.* While the vast majority of people you spray *will* be affected to some degree, the level of incapacitation will vary greatly from person to person. A subject sprayed with an atomized, conical spray pattern will also tend to react a bit differently than one sprayed with a ballistic stream pattern—the conical, mist-like spray is more easily inhaled deeper into the lungs than is the heavy liquid-based stream.

The effects of the ASR spray may be impeded if the sprayee is wearing sunglasses or other protective eye gear. Should an officer find himself the victim of a surprise ASR attack, he should try to avoid the spray by physically moving out of the way or placing something between the spray and his eyes and face.

Common Physiological Effects from ASR Spray

Eyes: Because OC is an eye irritant, you will generally observe tearing and redness of the eyes.

Respiratory System: OC, an inflammatory agent, may cause swelling of the mucous membranes lining the breathing passages. This will restrict the breathing to short, shallow breaths.

Ingestion: If swallowed, OC may induce a severe, burning, heartburn-like sensation or nausea.

Skin: May cause mild to severe burning sensation on exposed skin. May also cause redness. No lasting effects should be induced.

Other Effects You May Observe

- The subject's eyes may twitch or close tightly, or blink uncontrollably.[7]
- If the agent is swallowed or inhaled, the subject may cough or gag spasmodically[8] and feel like he is unable to breathe. This will redirect the subject's attention from fighting you but can also induce panic, which is a problem in its own right.
- The subject may rub his eyes, fall to the ground, or try to run toward you or away from the pain.
- The subject may continue to fight, become verbally abusive, or threaten revenge.
- The subject may beg for assistance and become 100-percent compliant.
- The subject may fold up into a ball of silence.

I have observed all of these effects exhibited by people in the field and in training. My experiences have also allowed me to observe the extreme ranges of these effects, from immediate, total incapacitation—usually because the person being sprayed was convinced that this is what would happen—to a few blinks and a shake of the head, after which the individual got more pissed off.

Having been sprayed directly several times with ASRs, I can also testify to the unpleasantness of the experience. However, as a result of my observations and personal exposures, I also know that I—as well as many potential adversaries—can continue to operate once sprayed, at least long enough to achieve a short-range objective, such as getting to the person holding the ASR.

The value of this knowledge in particular is one of the reasons I believe that anyone who carries an ASR should experience a full exposure to it while in training. When combined with a reality-based dynamic training scenario, during which you must fight to maintain control of your weapon and the situation, the benefits of being sprayed far outweigh the discomfort of having to endure the short-term effects of the agent.[9]

Learning how people react to and recover from exposure to the spray through experience and observation is also invaluable: it will aid you in detecting unusual symptoms or danger signs in subjects you have sprayed in the field.

Another reason that comes immediately to mind is the ability, after having been sprayed, to articulate with great clarity why you felt yourself to be in grievous fear when faced with a threatening adversary armed with an ASR.

It's also a rather pleasant experience to have a defense attorney, heavily armed with theatrical indignation at the severe, dehumanizing application of OC his client received, ask you if you realize just how terribly the defendant suffered.

"Yes, sir," you can reply calmly.

Should he then forget lawyer-tactics class 101 and ask a question to which he doesn't know the answer, such as, "Oh, really? And how do you know how terribly he suffered, officer?!" You can then out-theatrical him by remaining still for just one or two beats longer than necessary and then answering, "Because I've been sprayed directly in the face with the same agent."

"Okay, fine," I have had many officers say to me

after explaining my position on training with ASRs. "That's fine for you if you want to catch that shit with your face, but I'm not buying it." Then the main thrust of their argument will be driven home with, "After all, I'm carrying a gun that the department issued me, and I didn't need to get shot with *it* to know that it's gonna hurt."

Not a bad try, but unfortunately this argument doesn't hold up. The main reason being that, no, you're not shot with your firearm during training, but neither would you completely recover from a gunshot wound in 45 minutes.

The benefits of experiencing the effects of the spray firsthand under controlled conditions far outweigh the temporary discomfort you will be subjected to. It also tends to give you a great deal more respect for the amount of pain the agents do inflict when administered intentionally to a suspect, or unintentionally to an uninvolved person or co-worker under dynamic field conditions. This is important in our business because we are professionals and must be in every sense of the word.

In addition, we must also be prepared for those times when *we* are the recipients of an unexpected exposure to the spray, whether administered by a co-worker or an assailant.

FACING THE ASR

Another consideration that must be made is what to do should you be faced with someone who has his own ASR or someone who has managed to take yours from you.

The availability of ASRs on the open market makes the first possibility likely. The amount of time and training dedicated to ASR retention will have a decided effect on the second (see Chapter 15). Regardless of how the subject came into possession of the ASR, you must act decisively and quickly to avoid being sprayed and possibly disabled.

If possible, the first thing you need to do is get something between your face and the spray. Distance is preferred. Turning your head or holding something in front of your face (such as a clipboard) are options but will limit your ability to observe the subject.

After achieving enough distance to place you outside the range of the ASR (approximately 20 to 25

feet [6 to 7.6 meters] for most canisters), you should then order the subject to drop the ASR and call for backup. If you cannot retreat for any reason (e.g., space, injury), then you must determine just how much danger you are in based on the totality of the circumstances, your training, and experience. If you believe that the suspect intends to spray you (either because of his actions or words), that he has the ability to spray you, and that you will not be able to defend yourself or others once you have been sprayed, then you must act appropriately, using whatever level of force is required to control the situation.

Should you find yourself under attack by someone armed with an ASR and realize you have in fact been sprayed, you must do whatever is necessary to control your reactions and focus on defending yourself, retaining your personal weapons, and calling for backup. *This scenario must be prepared for during training!*

Personnel Decontamination

I can recall very clearly my first exposure to a full burst of ASR spray in a ballistic stream pattern. The experience can best be described as having liquid napalm liberally administered to your face. After the agent drops through the carrier, the burning sensation keeps increasing until it reaches a point where you think it can't possibly get worse—and then it does. The pain can be so intense and your focus on it so great that your involuntary breathing cycle may be interrupted. A common thought that goes through most people's minds after being sprayed is, "This really sucks!" The primary thought, however, is "MAKE IT STOP!"

The instructor who trained me and a dozen other troopers that fine March day nearly 10 years ago administered a good 1-to-2-second burst to each of us, outside, behind the state police academy, as we came at him aggressively, one at a time. After completing our attempt at attacking the gentleman, we were then guided to a half-dozen 5-gallon buckets filled with water.

As the pain increased, we thrust our heads into the buckets with greater and greater urgency, for underwater you could open your eyes and the pain would be abated for as long as your topknot was submerged. Unfortunately, there weren't enough buckets to go around. Looking at it in retrospect, I'm glad that none of us were armed.

Paper towels were also on hand, and we received directions to "blot" the agent off each other's faces in between goes at the buckets. Unbeknown to us at the time, however, the oily OC resin would float on top of the water in the buckets after a contaminated head was removed, and the next time a face was driven into the cold liquid, a fresh coat of agent would be administered.

All in all, this was an interesting training experience, punctuated with much sadistic and masochistic laughter, as well as chants of, "Sir, I love my bucket, Sir."

While this decontamination method might have produced an interesting experience, it proved to be far from efficient or desirable—*especially* since it occurred in a controlled training environment. In addition, a goatfuck as described above is not conducive to convincing other members of the department as to the viability of being sprayed with the agent—especially if being sprayed is not required in order to be issued the ASR.

Fortunately, other decontamination options are now available. Products such as Cool It!, Bioshield, Sudecon, and Decon Aftercare can be applied directly to contaminated areas immediately after exposure, greatly reducing both discomfort and recovery time during training and in the field. These products, generally available as a spray or in towelette form, work particularly well with straight OC formulations but are also effective with most ASR combination formulations.[10]

If these products are not available, you can still decontaminate yourself or other individuals efficiently using cool soapy water (mild soap), paper towels, and large amounts of clear water.

The cool soapy water can be sprayed onto the face with a common plastic pump bottle. The mixture should be allowed to run freely off the face.

Paper towels, if used, should be pressed gently against the affected area and then immediately removed and discarded. In this way the agent is removed from the surface of the skin, not ground deeper into the pores.

The eyes and face should be continuously

flushed with large amounts of fresh, clear water. Saline solution applied directly to the eyes is also very effective. The eyes should be forced open as soon as possible, and the face should be turned into the wind, allowing both the affected subject's tears and the air to assist in clearing the agent.

IMPORTANT: *If you or the subject you are decontaminating is wearing contact lenses, they must be removed as soon as possible. If you have not been trained to remove another person's contact lenses and an individual is unable to remove his own, call for medical assistance immediately.*

The nose should be kept clear by blowing to remove discharge and agent particulates. Breathing should be controlled as much as possible in order to prevent hyperventilation.

The skin should be flushed with large amounts of cool, clear water. No salves, creams, lotions, or oils should be applied.

Showering with cool water and mild shampoo will also aid in decontamination. Just be advised that when you shower, some of the agent particulates may be reactivated, at which point you may develop a whole new appreciation for the burning effects of capsaicin.

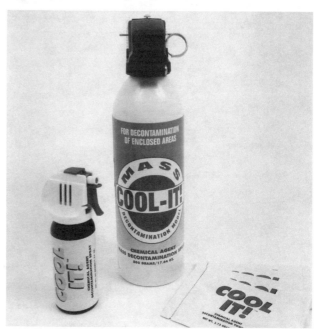

Cool It!

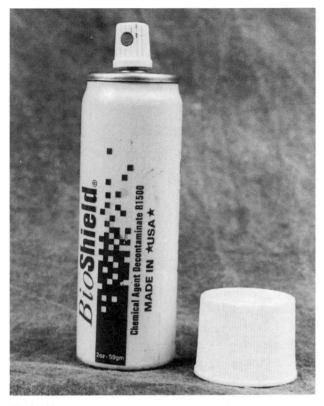

Bioshield.

Decon Aftercare.

Sudecon.

Area and Equipment Decontamination

One of the most attractive aspects of using an OC-based agent is the minimal amount of time and effort required to decontaminate personnel, equipment, and areas. The vast majority of decontamination can be accomplished simply by washing away any remaining OC agent with soap and water and allowing the area or equipment to ventilate. The effects should be completely gone within an hour.

When washing clothing that has been contaminated, wash them separately in cold water with detergent. It is also recommended that the clothing be air-dried as opposed to being placed in a hot-air dryer, for this could reactivate any particulates that may still be present.

Positional Asphyxia

Positional asphyxia. These two words represent a phenomenon that has impacted greatly on the way police operated in the United States and abroad during the past decade. Many departments and agencies are still just starting to address this issue in training and through written policies and procedures.

Certainly, this concern is warranted: we are absolutely responsible for the care and safety of anyone we take into custody. Some recent research on positional asphyxia, however, indicates that much like two other words from the last millennium that had great impact on law enforcement in the United States—"Rodney King"—the term *positional asphyxia* and the controversy surrounding it and its link to chemical spray use may be more hype than substance.

Specifically, positional asphyxia has been identified as a phenomenon that results when the position of an individual's body interferes with his respiration, resulting in asphyxia (suffocation). The "hog-tied" position in particular has been associated with several in-custody deaths. Restraining an individual by tightly strapping him or placing him in the prone position is also believed to have the potential to compromise an individual's respiratory system.

What is often overlooked, however, is that other syndromes can also contribute to sudden in-custody deaths. Cocaine-induced excited delirium (CIED) is one such syndrome. Individuals afflicted with this syndrome exhibit similar behaviors, including bizarre or aggressive behavior,[11] shouting, paranoia, panic, violence toward others, unexpected physical strength, and sudden tranquility.[12] Sudden death occurring either during or immediately after a person exhibiting signs of CIED has been involved in a violent struggle has been documented, though autopsy findings are generally nonspecific.

Neuroleptic malignant syndrome (NMS) is another syndrome that can cause sudden, unexpected death. Individuals suffering from NMS may exhibit similar behaviors as someone afflicted with CIED. NMS generally occurs in psychiatric patients who are taking antipsychotic medication. Some researchers believe that this condition, though poorly understood, may be related to a cardiac event due to psychological stress. In other words, someone suffering from NMS could literally be scared to death. Again, autopsy findings have generally been inconclusive, but anyone exhibiting any of the behaviors described above must be closely monitored and should be taken immediately to a medical facility for evaluation.

As respiration depends upon three critical elements—i.e., (1) the gas exchange function of the lungs, (2) the unobstructed flow of oxygen through the airway, and (3) the muscular pump or bellows function that ventilates the lungs[13]—anything that interferes with any of these functions could be construed as a contributing factor to a resulting death.

So while the position a person is restrained in may impact on his ability to breathe, other significant contributing factors should not be ignored if an in-custody fatality occurs. Some of these additional contributing factors include (but are not limited to) obesity, intoxicants, physical conditioning, prior existing medical conditions, and involvement in a struggle or other violent, high-exertion activity.

CARE AND MAINTENANCE

ASR canisters should be kept clean and inspected routinely. The nozzle, in particular, should be checked for dirt, dust, or clogging. Any signs of damage, erosion, or leakage to any part of the unit should be immediately reported to the departmental armorer or chemical agent officer.

Once a week, the canister should be removed from the pouch and shaken to ensure that the agent/carrier formulation stays mixed. The canister should also be test-sprayed occasionally to ensure that the propellant has not leached out through the canister seal. Should this happen, the canister may feel "full," yet no liquid will be projected out of the nozzle. The time to discover that this has happened is not when you are attempting to spray a subject in the field.

Canisters should not be left in motor vehicles or otherwise left exposed to extreme heat or cold. Any aerosol canister may burst if exposed to temperatures of 120°F (48.8°C) for prolonged periods. Temperatures below 32°F (0°C) may also adversely affect an ASR, causing it to discharge slowly or improperly when used.

Some manufacturers recommend that OC canisters be kept upside down when placed in long-term storage. It is believed that this keeps the seal wet so it does not dry out and become porous.

MORE ABOUT OLEORESIN CAPSICUM

As a result of its current (and undoubtedly continuing) popularity, a closer look at the OC agent is warranted. As stated previously, there are many misconceptions and half-truths about this agent that continue to circulate throughout the law enforcement community. One of the more prevalent misconceptions relates to the rating of the "strengths" of the different formulations. These strengths are commonly indicated by one of three methods:

- Scoville heat units (SHUs)
- The percentage of OC contained within the product solution (e.g., 5-percent OC)
- The capsaicinoid concentration

The various manufacturers of OC sprays use each of these rating systems. Depending upon how they are used, they can provide either an excellent indicator of product strength or a misleading marketing gimmick.

Scoville Heat Units

A panel of five "tasters" originally established the SHU system of measurement.[14] These five individuals would literally taste samples of various products and then rate them according to the heat or burning sensation they caused on the tongue. This methodology, while hardly objective or scientific, is commonly used by chefs and others in the food preparation services to rate the measurement of "hotness" or heat in food products. SHUs are usually expressed in millions (e.g., 2 million SHUs).

Percentage of OC

The percentage of OC in any given formulation is often used to indicate the product's strength. This system of measurement, however, is not an accurate indicator of the product's actual strength, because the strength of the OC *itself* varies from batch to batch. The reasons for this variation range from the quality of the peppers used to the way the OC was extracted to other chemicals with which it was blended. What this means is that it is possible for an ASR containing 5 percent of an extremely hot batch of OC to be just as potent as one containing 10 percent of a weaker batch.

As for severity of effects, given two batches of equal strength, the only significant difference in performance between a formulation containing 5-percent OC and one containing 10-percent OC is

the amount of time needed to recover from the application. Generally speaking, a shorter recovery time is more desirable from a law enforcement perspective.

Capsaicinoid Concentration

Measuring the capsaicinoid concentration is considered the best indicator of the actual pungency, or strength, of the OC formulation.

Capsaicinoids are the active components in OC. It is the capsaicinoid compounds themselves that cause the actual burning sensation. The three main compounds that are classified as capsaicinoids are as follows:

1. Capsaicin
2. Dihydrocapsaicin
3. Nordihydrocapsaicin

What the capsaicinoid concentration indicates is the amount of these active compounds within a solution. This known quantity can then be used to gauge the actual strength of the OC solution. This method is considered far superior to the two listed above for estimating the strength of any given OC formulation because a 0.2-percent capsaicinoid concentration in one canister will produce the same degree of hotness as a 0.2-percent capsaicinoid concentration in another.

OC—The Natural Choice?

It is often argued that OC is neither exceedingly harmful nor dangerous because it is "natural." And it *is* natural in that the agent capsaicum is derived from various types of chili peppers.

However, as Bert DuVernay, director of the Smith & Wesson Academy once noted, a 4-foot long yellow-eastern pine 2 x 4, being made of wood, is also all natural—but you wouldn't want to be hit over the head with one. In fact, to carry this analogy even further, you could also argue that like capsaicum, lead is a natural substance.

Of course, both substances lose a good deal of their "naturalness" when projected at a human being at high velocity.

The point I am trying to make here is that any weapon can be dangerous if used improperly. Even when used properly, chemical agent weapons may still prove dangerous or even deadly given the right circumstances.

ASR Holders or Carriers

ASRs can be carried as free units in pockets, briefcases, glove boxes, or gear bags. Regardless of how you carry any ASR, you must remember that you are responsible for its use or misuse should it fall into unauthorized or untrained hands. If you do carry an ASR without keeping it in some type of protective pouch or container, then you should choose a model with some type of integral safety cap, latch, or pin that prevents the unit from discharging unintentionally. Otherwise you run the risk of contaminating yourself, others, clothing, vehicles, equipment, etc.

Some ASRs produced for duty use come equipped with some type of pouch or holder. If a pouch or holder is not included with your ASR, then purchase one that will allow the ASR to be carried securely and accessed quickly. A few examples of ASR carry devices are shown below.

ASR carry devices, left to right; molded plastic, nylon, and leather. Most use some type of snap-closure device to keep the units secured. Some use Velcro-like fasteners. In my experience, the Velcro-like fasteners tend to be less effective, especially over the long term. It should also be noted that leather cannot be sterilized. Should your leather gear be exposed to blood-born pathogens, it must be destroyed!

NOTES

1. Doug Lamb, *Pepper Sprays: Practical Self-Defense for Anyone, Anywhere* (Boulder, Colo.: Paladin Press, 1994), viii.
2. "Gas" actually being another misnomer, as CS and CN agents are actually finely ground crystal-like particles.
3. While assigned to MSP Special Operations, I spent two years working with Tom Robbins and Tim Donnelly in the densely populated city of Chelsea, Massachusetts (population approximately 14,745 people per square mile). Part of the federal Weed and Seed initiative, the Chelsea Detail served as a pilot program for Massachusetts State Police involvement in high-crime-area community policing operations.
4. A DuPont product.
5. The Thomas A. Swift (TASER) electric rifle fires two metal barbs at high velocity. These barbs are tethered to the hand-held unit by thin wires. Once the barbs are attached to the intended subject, a trigger is pressed and a burst charge of approximately 50,000 volts of current is administered through the wires into the subject.
6. See "Point Shooting Technique" in the Glossary.
7. This severe twitching or spasmodic contraction of the eyes is referred to as *blepharospasm*.
8. This is believed to be a response to bronchoconstriction, a constriction of the airway.
9. Instructor Phil Messina of Modern Warrior compiled some interesting statistics regarding this type of integrated training. He found that when involved in a struggle for the pistol after being sprayed with OC, individuals who were experiencing their first exposure to the agent would lose control of their sidearm approximately 90 percent of the time. On the other hand, those who had been sprayed prior to the exercise kept control of their pistols 70 percent of the time.
10. As an aside, some trainers have discovered that applying a product such as Decon Aftercare to the face and nostril area *prior* to being sprayed greatly decreases the effects and decontamination time.
11. Interestingly, aggression toward objects, particularly glass, has been documented.
12. J. Granfield, and Petty C. Onnen, M.D., "Pepper Spray and In-Custody Deaths," *Science and Technology* (March 1994).
13. Donald T. Reay, M.D.; Corinne L., Flinger, M.D.; Allen D. Shilwell, M.D.; and Judy Arnold, "Positional Asphyxia During Law Enforcement Transport," *The American Journal of Medicine and Pathology*, 12(2) (1992): 90–97.
14. *American Spice Trade Association*, Analytical Methods 21.0.

Fog Generators

In use by civil law enforcement since 1968, fog generators are produced in many sizes, shapes, and configurations. They are designed to dispense oil-based insecticides, fungicides, germicides, disinfectants, odor control sprays, and other chemical products easily, effectively, and economically.

One of the major manufacturers of fog generators is Curtis Dyna-Fog, Ltd., based in Westfield, Indiana. Russell Curtis, the founder of what would eventually become Curtis Dyna-Fog, Ltd., was first exposed to the idea of a Pulse-Jet engine at Wright Airfield (now Wright-Patterson Air Force Base) in Dayton, Ohio, immediately after World War II. The Pulse-Jet had been the power source for the German VI rockets during the war and would later become the foundation on which Curtis would build his company. Today's modern thermal fog generators still employ the same basic technology.

The Golden Eagle MK XII-D pepper fog generator. At just over 4 feet in length and weighing approximately 27 pounds (10 kg) when filled, this unit is both highly maneuverable and manageable.

THERMAL FOG GENERATORS

Thermal fog generators produce large volumes of hot gases that flow at high velocity. When a liquid formulation of CN, CS, OC, or smoke is introduced into these high-velocity gases, it is instantly atomized and then condensed. This produces dense clouds of irritant fog or inert smoke.

The size of the particles within this fog are controllable and can range from 0.5 to 50 microns or more. The size of the particles produced depends on the specific formulation and flow rates that are used.

Both the electric-start Golden Eagle pepper fog generator and the older manual pump-start GOEC[1] unit shown on these pages operate using the thermal aerosol principle. Both units also possess resonant pulse-jet engines that are fueled by gasoline.[2]

Smaller, more portable fog generators that employ standard piston engines are also produced in various configurations, including a special backpack version for small scale applications.

Thermal fog generators such as these are extremely efficient for police chemical agent operations, both tactically and economically. For example, one minute of continuous CS fogging, on average, will put out as much irritant as ten pyrotechnic CS hand grenades. In addition, the operator can instantly vary the concentration of the agent that is released, providing him a greater degree of control and flexibility when dealing with the dynamic realities of civil disturbances and (for corrections personnel) large-scale inmate control situations.

IMPORTANT: *The fogger units shown in this section are intended only for outdoor use or in enclosed areas with volumes of more than 500 cubic feet (14 cubic meters). Use of these machines in confined areas of less than 500 cubic feet presents a serious hazard from fire or explosion.*

How They Work

The engine is essentially a tube with a combustion chamber, an intake valve, and a supply of combustible mixture of fuel and air (see illustration below). Fuel is drawn from the fuel tank and forced into the antechamber and combustion chamber where it is ignited by a spark plug. An explosion then occurs in the combustion chamber, driving the gases out of the engine exhaust tube. The negative pressure created by the gas flow out of the engine tube causes the intake valves to open. This allows more air to pass through the venturi of the carburetor. The air passing through the carburetor aspirates fuel from the carburetor in a combustible mixture. This mixture is ignited again, and the cycle is repeated. An integral glow coil located in the engine tube sustains repeated

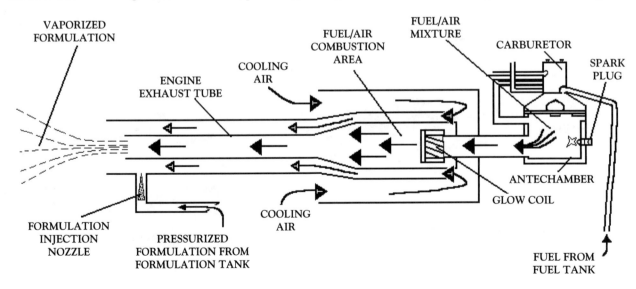

Illustration 12-1. Thermal fogger generator.

ignition cycles. The frequency of repeated explosions is many times per second.

Once the engine is running smoothly, the operator can then depress the formulation on/off button, opening the formulation valve. The liquid formulation is then delivered to the engine tube through the formulation injection nozzle and injected into the high-velocity pulsating flow of hot gases. The formulation is then broken down by the pulsating gases and discharged into the atmosphere as a fine white particle fog.

Fog Ignition: A Fire Hazard

Under some conditions, it is possible for the fog being discharged to ignite, creating a torching effect. This is more likely to happen when using a "wet fog" setting than a "dry fog" setting because there will be more combustible material in the air. Though the business end of the engine exhaust tube has been designed to limit the possibility of this occurring, anyone who operates these machines must be aware of the hazard and know how to avoid it, or, should they encounter it, how to control it.

Formulation ignition can be caused two ways.

Ignition from an External Source

Ignition can be caused by a flame or spark that occurs 6 to 8 inches (5.2 to 7.6 centimeters) from the end of the discharge tube. This should be considered if fogging in buildings, as electrical sparking, arcing, pilot lights, or other open flames might be present. Although the operator may use care and common sense when fogging indoors, the dense fog will still limit his vision. That is why it is recommended that, if possible, all pilot lights be extinguished and power turned off within the objective area before employing the fogger indoors.

Ignition from an Internal Source

It is also possible for the machine itself to ignite the formulation under certain conditions. If the engine stops running (either because it is out of fuel or experiencing mechanical malfunction), it is possible for some of the formulation to flow into the hot engine tube where it will vaporize. Some of this vapor may then be drawn back to the red-hot engine

combustion chamber, where it can ignite. Should this occur, the ignited vapor may flash back through the discharge end of the engine tube, igniting any formulation flowing through the tube.

Fog Ignition—Immediate Action Drill

Regardless of whether the ignition is caused by an external or internal source, the operator must quickly release the formulation on/off button to stop the formulation flow and fog discharge. You must also take precautions to avoid allowing any burning formulation from coming into contact with any flammable substances or materials, as fire may result. Also be aware that under certain conditions, if a fire occurs, the possibility of explosion will exist.

It should be noted that the possibility for either fire or explosion can be limited if safety precautions are taken when using thermal foggers and if the machines are regularly serviced and properly maintained.

THERMAL FOGGERS—SAFETY PRECAUTIONS

1. Protective clothing and ear and eye protection should be used when operating thermal fog generators. Respirators must be used when employing irritant formulations.
2. Thermal foggers are fueled by gasoline. When hot, the fogger's engine, glow coil, and engine exhaust tube can easily ignite either the fuel or fuel vapors.
3. Fogging formulations can also ignite. This is true whether they are in liquid or vapor form. That is why you should extinguish all oil and gas pilot lights and turn off all electrical power before using the fogger indoors. You must also avoid fogging near open flames, cigarettes, or other hot materials. *If fog ignition occurs or the engine quits for any reason, the operator must immediately stop the fog discharge by releasing the formulation on/off button!*
4. Never use a wet fog in a closed area. (To determine if the fog is wet, pass a piece of dark paper or a shiny object through the fog approximately 2 feet (61 centimeters) in front of the discharge tube. If any accumulation is visible on the paper or object, this means that the fog is

wet and the formulation metering valve should be changed to a dryer setting).

5. Do not fog any enclosed space of less than 500 cubic feet (14 cubic meters).

6. Never wedge, tape, or block open the formulation on/off button or leave the machine running unattended.

7. Never look into the engine discharge tube when the machine is running.

8. Never place the machine on its side when the fuel or formulation tanks contain liquid.

9. Always maintain a distance of at least 2 feet (61 centimeters) between the engine discharge tube and external objects. This includes walls, furniture, vehicles, and other objects. Blocking the discharge tube will cause the unit to overheat, possibly resulting in permanent damage to the machine, fire, or an explosion.

10. Keep the fogger moving. Leaving the fogger pointed at the same area or object for too long can result in flammable substance buildup.

11. Never operate a machine that has been damaged! Fire or explosion may result.

12. Avoid using thermal foggers in windy conditions. Not only may the fog be blown in directions other than that intended, but fog ignition can result if the engine stops and vaporized formulation is blown into the combustion chamber.

13. Do not allow anyone to enter or play in the fog. Children in particular may be attracted to the clouds of inert fog; visibility is limited in a fog, and injury may result.

14. Foggers must be properly serviced, maintained, and stored to ensure optimal operation.

Preparation for Use

When preparing any thermal fog generator for use, you should follow all manufacturer's instructions and recommendations. Certain precautions should be taken:

- The generator should be placed in an open, uncluttered, and well-ventilated area.
- Care should be taken to ensure that there are no open flames or flammable materials in the immediate area (including lit cigars and cigarettes).

- The machine should be placed on a stable workbench or on a concrete pad.
- Clear, level ground will suffice if you are in the field.

After taking the above precautions, you are now ready to proceed. To start the generator, take the following steps:

1. Fill the fuel tank with clean gasoline.[3] Once filled, ensure that the fuel cap is tight and then wipe off any spilled gasoline from the machine. You should then wait an appropriate amount of time to allow any unseen spills to evaporate.

2. Fill the formulation tank with an appropriate amount of the desired agent or smoke formulation.[4] You should only use enough formulation to complete the specified mission.[5] This will ensure that the formulation tank is empty once the mission has been completed.

IMPORTANT: *Never use any substances from unmarked containers or from containers with altered or questionable labels.*

3. Start the resonant pulse jet engine.

Starting the Golden Eagle MK XII-D Electric Start Model

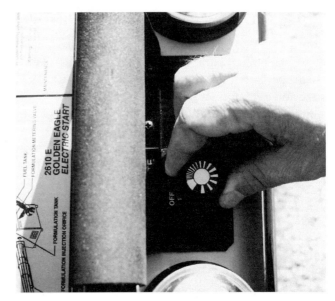

1. Ensure that the formulation metering valve is set to "0" (fully counterclockwise).

2. Lift the ON/OFF control knob and place it in the ON position. (**IMPORTANT:** Do Not depress the formulation ON/OFF button until ready to fog.)

3. Press the primer bulb until fuel appears.

4. Simultaneously depress the ignition and air switches. Once the engine catches, release the air switch but continue to depress the ignition switch until the engine is running smoothly.

Starting the GOEC Mark XII-A Pump Start Model

Bert DuVernay attempting to start an older-model General Ordinance Equipment Corporation (GOEC) Mark HXII-A Pepper Fog generator using a hand pump.

Bert DuVernay fogging with an older-model GOEC Pepper Fog generator. The hand pump–equipped models can be difficult to start, especially in colder weather. Bert has no problem getting this unit to operate even in frigid temperatures, in great part because he carefully maintains it and operates it often. Bert recommends starting and running the fogger every one to three months.

FOGGING

Once the engine is running smoothly, depress the formulation control valve (on/off button) to fog. Adjust the formulation metering valve to achieve the desired fog quality. A "dry" fog is desirable, especially when fogging in a closed area. To determine if the fog is wet, pass a piece of dark paper or a shiny object through the fog approximately 2 feet (61 centimeters) in front of the discharge tube. If any accumulation is visible on the paper or object, the fog is wet and the formulation metering valve should be changed to a dryer setting.

To Stop Fogging

To temporarily stop fogging, release the formulation on/off button.

To completely stop fogging after the mission is completed, do the following:

1. Release the formulation on/off button.
2. Rotate the formulation metering valve to "zero."
3. Stop the engine by placing the on/off control knob into the off position.
4. Drain any unused formulation from the formulation tank into its original container for storage.

CARE AND MAINTENANCE

It is imperative that the machine be properly maintained. If formulation or formulation residue is allowed to remain in the formulation systems after use, then frozen valves and plugged lines will result. *This can begin to occur within one hour after the machine is shut off.*

After Each Use

Immediately after each use you must drain the formulation tank and then flush out the entire formulation system with flush mix. To flush out the system, do the following:

1. Pour the recommended amount of flush mix (usually 1 pint) into the formulation tank and slosh it around thoroughly. Replace formulation tank cap.
2. Start the engine and fog out all the liquid (flush mix and residue) in the formulation tank.
3. Examine the formulation tank, lines, and valves for formulation buildup. If formulation buildup is detected, increase the amount of flush mix used.

IMPORTANT: *All precautions and safe operating procedures for fogging apply when flushing the system.*

Additional Maintenance Procedures

Following regularly scheduled maintenance procedures will ensure that the fogger gives your department years of dependable service. It will also assure you that the machine will be ready to go when you need to respond to a quickly developing situation.

In addition to flushing the formulation tank and system thoroughly after each use, *after every four hours* of use you should also do the following:

1. Clean the engine discharge tube. The cleanout brush is inserted completely into the tube while you rotate it in a clockwise direction. Then pull the brush back out of the tube, again rotating it clockwise. Loose carbon will be removed from the tube the next time the engine is started.

Cleaning the discharge tube.

2. Clean the formulation filter located in the tank neck with detergent and water;[6] then thoroughly dry and reinstall it. Do not fog without a formulation filter because damage to the system will result.
3. The engine neck, spark plug, and control linkages should also be checked for wear and cleaned and adjusted as required.

After every 12 hours of operation you should do the following:

1. Clean formulation injection nozzle.
2. Check the fuel filter and replace if necessary.
3. Clean the formulation injection orifice assembly.
4. Check the batteries (electric start machines) and replace as needed.

Storage
Unless used periodically for missions or training purposes, the fog generator will need to be properly stored. One recommended method is provided here.

1. After properly flushing out the formulation system as described in the "After Each Use" section, remove the formulation tank cap and the drain plug located on the bottom of the formulation tank, rinse the inside with a flushing agent, and drain. Then replace the drain plug and tank cap.
2. Remove the fuel tank cap and the fuel tank plug and carefully drain all fuel from the tank. After the fuel tank is empty, replace the fuel tank drain plug and cap.
3. Press and hold down the ignition switch while simultaneously pumping the primer bulb until no further firing occurs and no more fuel is observed in the engine's antechamber. This will clear any remaining fuel from the fuel lines and fuel valve.
4. The batteries should then be removed and stored separately in a cool, dry place.
5. The machine itself should be stored in a cool, dry place, preferably in its original shipping carton. If the carton is not available, then the machine should be placed off the ground and covered to keep dust and dirt from accumulating.

NOTES

1. The General Ordnance Equipment Corporation is defunct.
2. The exhaust of a pulse jet engine is low in pollutants. This is because of the extremely high efficiency of the engine, which actually burns all combustible material to nonpollutant end products.
3. The manufacturer recommends that fresh, clean gasoline with a minimum octane rating of 87 be used. Either regular or unleaded fuel is acceptable.
4. Available formulations include CN, CS, OC, Smoke, and Flush Mix. Containers vary in size from pints (CN, CS), to quarts (CN, CS, OC, Smoke, Flush Mix), to gallons (Smoke).
5. The amount of formulation required to fog a specific area depends on several factors, including the viscosity of the formulation, the formulation tank pressure, and the operating characteristics of the engine. You must read and be familiar with all safety precautions and operating instructions that apply to the specific fog generator and formulation you are using.
6. Automotive carburetor cleaner can be used for heavy deposits.

Compressed Gas Projector Systems

Compressed gas projector systems look like and function similarly to the common fire extinguisher. Agent material in powder or liquid form is loaded into the metal body or cylinder. The body is then sealed, and the tank is pressurized with (most commonly) nitrogen through a valve. Once the recommended pressure is reached, the unit is ready to be used. Agent is released from the unit when a valve is opened. Once depleted of agent and/or propellant gas, the unit can be refilled. This feature makes compressed gas projector systems extremely cost efficient.

Operator testing the ISPRA Model-5 Protectojet.

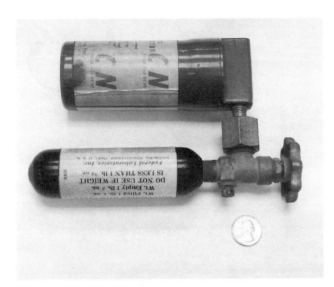

Federal Laboratories Tear-Gas Dust Projector. These small, portable devices were designed to shoot CN tear gas dust into areas through small openings. These types of units, while simple to produce and use, are not commonly employed today, having been replaced by such products as Aerko's specially modified Clear Out grenades (See Chapter 7). The U.S. quarter is shown for scale. (The tear gas projector is from S&W Academy collection.)

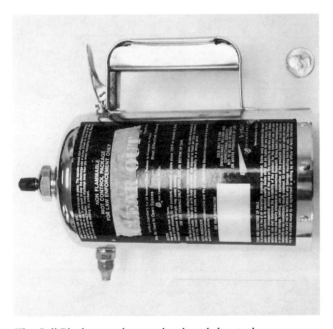

The Cell Blocker is a chrome-plated-steel chemical agent projector. It uses compressed air to propel up to one quart of chemical agent in liquid carriers out to a range of approximately 19 to 25 feet (5.7 to 7.6 meters), depending upon elevation of nozzle. When the release lever is snapped in place, this projector will expel its contents for approximately 28 seconds. (Cell Blocker is from the S&W Academy collection.)

Many compressed gas projectors are similar in appearance and function to common fire extinguishers. Units like the Defense Technology MK-21 and MK-46 high-volume-output OC aerosol projectors shown here are very simple to use. These units are loaded with a formulation of OC in a liquid state. The refillable aluminum dispenser body is pressurized with nitrogen gas. When the safety pin is removed, a simple press of the actuator handle allows the pressurized agent to be emitted downrange in the form of a high-volume ballistic stream. The MK-21 holds approximately 21 ounces of agent formulation and is pressurized with nitrogen to 190 psi (at 70°F) when fully loaded. It boasts an effective range of 25 to 30 feet (7.6 to 9.1 meters) in stable air conditions. The MK-46 holds approximately 46 ounces of agent formulation pressurized with nitrogen at 190 psi (at 70°F). The MK-46 has an effective range approximately the same as the smaller MK-21, but allows more half-second bursts (approximately 12 to14 half-second bursts for the MK-21, 24 to 26 for the MK-46).

The nitrogen-pressurized GTP-1 (top) offered by Guardian Protective Devices is configured with an efficient actuator valve that allows for easy one-handed operation. When the actuator handle is pressed, the pressurized agent is emitted in the form of a high-volume ballistic stream. Operating the Model-5 Protectojet from ISPRA (bottom) with one hand can be done, but it is awkward. Larger units like these can have an effective range of up to 60 meters (196.8 feet) when environmental conditions are right.

ISPRA Model-5 Protectojet valve (top) and Guardian Protective Devices GTP-1 valve (bottom). Both are equipped with safety pull-pins. The GTP-1 is also equipped with a pressure indicator gauge and can be operated easily with one hand when a shoulder strap is used.

USE

Larger units such as the ISPRA Model-5 Protectojet are configured to allow easy use when the operator is mobile. The shoulder strap and tank cover shown above permit the unit to be carried and maneuvered comfortably. The Protectojet valve configuration requires two-handed operation for best results.

Compressed-gas projector systems are generally used to clear or neutralize large areas as opposed to being used against lone individuals.

The agent should be released in half-second bursts as needed. Units tend to deplete fairly quickly because of the high volume of pressurized agent released during each burst. The stream or high-pressure cloud should never be aimed directly at anyone's eyes because severe injury can result. This danger increases as the range between the unit and individuals being sprayed decreases.

CARE AND MAINTENANCE

When empty, these devices should be completely depleted of all agent and propellant and then reloaded or sent to the company or other authorized service location to be refilled. If the units are shipped from the service location back to your agency or department, you will need to get them pressurized because the Department of Transportation does not allow shipping of pressurized units. When refilling or pressurizing the unit yourself, all manufacturer-recommended procedures should be followed. Pressure levels should never be exceeded—canisters can burst and valves can be ruptured, possibly damaging equipment and causing injury to personnel.

Storage

Units should be stored in a cool, dry place, preferably in their original shipping cartons. If the carton is not available, then the projectors should be stored so that they are protected from impact or puncture. The valves must also be protected from damage or impact as these units are pressurized and serious injury or destruction of property could occur. Nozzles should be kept clean and valves checked periodically to ensure that there is no leakage.

Gas Launchers

Numerous companies throughout the world produce gas launchers in a variety of different sizes, shapes, and configurations. Whether manufactured in the West or the East, however, their primary purpose is to deliver the smoke, chemical agent weapon, or specialty impact munition to the objective area by launching it through the air. This allows the operator to maintain a greater distance from the objective area while he introduces the agent or munition precisely where needed.

At closer ranges, chemical agents in micropulverized form can be expelled directly out of the end of the muzzle of a gas launcher and into the atmosphere (see Chapter 9).

Canister-contained agents may also be delivered to an objective area when the canister itself is launched either out of the muzzle of the launcher or out of an adapter cup designed for this purpose.

The great interest in less-lethal weapons and gas launchers in particular over the past few decades has led to the development of some of the best designed and most accurate launchers ever produced. Many of these modern launchers have rifled bores, which, when used with specially designed munitions, will produce consistently accurate shot-placement.

Although there are exceptions, the vast majority of handheld gas launchers are produced in 37, 38, or 40mm calibers. A few examples of various launcher designs and some of their configurations are illustrated in the following pages.

SUBCALIBER WEAPONS USED AS LAUNCHERS

Revolver

S&W Model 65, .357 Magnum pistol equipped with launching adapter. A special cartridge is used to launch the small percussion mechanical fuze-equipped pyrotechnic grenade up to 120 meters (131 yards). (Launcher from the author's collection.)

A .38 or .357 magnum revolver may be used to launch small chemical agent weapons or smoke grenades. The adapter shown above is designed to fit on the end of the barrel of a Smith & Wesson or similar revolver. These particular adapters and the unique Mighty Midget percussion mechanical-fuzed munitions used in them are not currently being

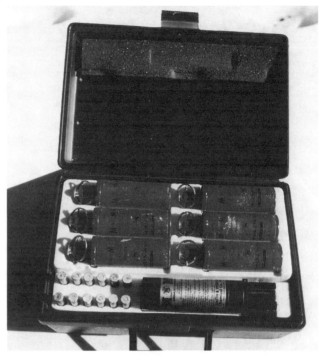

The S&W Mighty Midget Pistol Launching Adapter Kit. These particular munitions are no longer produced in the United States, though many may still be found in police department inventories. These and any other expired munitions should be properly disposed of. (Launching kit from the author's collection.)

manufactured in the United States. A similar unit (designated the .38 M1-CEV Revolver Launcher) that does employ the percussion mechanical-fuzed munitions is currently being produced by the Brazilian firm Companhia de Explosivos Valparaiba (CEV). Other launching adaptors (e.g., the DEF-TEC No. 99) that allow the revolver to be used to launch smaller or "tactical-sized" versions of standard M210A1 mechanical fuze-equipped grenades (such as the DEF-TEC No. 98) have also been produced.

Semiautomatic Pistol

Semiautomatic pistols are not used extensively as gas or grenade launchers, although they can be adapted for the purpose. China North Industries Corporation[1] produces a series of 35mm Anti-Riot Pistol Grenades that can be fired by various types of 7.62mm pistols and submachine guns. The 35mm munitions, which include explosive and

Illustration 14-1. Drawing shows configuration of 7.62mm semiautomatic pistol loaded with 35mm Anti-Riot Pistol Grenade. The special blank cartridge is used to launch grenades to ranges of up to 110 meters (120 yards).

nonexplosive tear gas rounds, as well as kinetic energy[2] and dye marker[3] rounds, have thin tail units attached to the nonmetallic grenade body that slip inside the weapon's bore. The grenades are launched from the weapon using a special blank launching cartridge (see Illustration 14-1).

SUBMACHINE GUNS USED AS LAUNCHERS

Submachine guns, such as the Israel Military Industries' UZI shown in the photo below, can also be used as gas launchers. The grenade shown in this

photo has been designed with a bullet trap in the tail unit. This allows it to be launched by a standard ball cartridge, eliminating the need for changing magazines or the configuration of the weapon.

Other grenades designed to be launched from submachine guns (e.g., the M6 produced by SIMAD Stacchini in Italy) require the use of special launching cartridges and firing adaptors.

Regardless of how they are configured, grenades of this type will generally have a range of up to 100 meters (109 yards) depending on the submachine gun and ammunition used.

LONG GUNS USED AS LAUNCHERS

Rifle

Illustration 14-2. Muzzle-launched grenade.

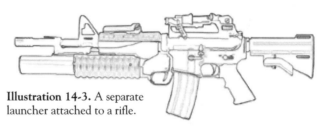

Illustration 14-3. A separate launcher attached to a rifle.

Rifles of all types and configurations have been adapted for use as grenade launchers. Depending on the design of the grenade, special adapters may or may not be needed. Grenades can be muzzle-launched (Illustration 14-2) to ranges up to 150 yards (137 meters), depending upon the type and weight of the grenade.

Attaching a separate launcher directly to the rifle is another option (Illustration 14-3). The M203 grenade launcher is shown here mounted on a Colt M16A2 (M4) Carbine. Integral launchers such as this are available in both 40mm and 37/38mm. The M203 incorporates a self-cocking firing mechanism that allows the launcher to be operated as an independent weapon even though it is fixed to the rifle. Munitions can be launched with a fair degree of accuracy out to ranges of 350 yards (320 meters).

Shotgun

Bert DuVernay instructing operators in the two-man team concept. First, the loader inserts CAW grenade into launching cup and removes the pin. . . .

. . . The loader then hands the firer a special launching round that the firer loads into the shotgun's breech. This is done for safety's sake. . . .

. . . The loader, standing directly behind the firer, then directs the firer to adjust windage and elevation. In addition to verbal commands, this is accomplished by having the loader press on the firer's left or right should to indicate windage, and by pushing or pulling him gently back and forth to indicate elevation. This is done to prepare for the possibility of having to operate in extremely loud or confusing environments as will be found in civil disturbance situations. When ready, the loader gently pushes firer's head forward to indicate he should fire, as well as to shield his eyes from any propellant or sparks. The loader does the same. At this point the firer presses the trigger and launches the CAW grenade.

With the addition of a launching cup and the use of special launching rounds, a standard 12-gauge shotgun can be used to launch chemical agent and smoke canisters approximately 90 yards (82 meters) into an objective area.

Care must be taken when employing the shotgun in this manner. It is especially important that operators be trained properly in the use of the shotgun as a launcher because the familiar pump gun truly becomes a different animal when so equipped. It is also recommended that operators be organized and trained to function as two-man teams, consisting of a firer and a loader.

The firer on the team controls the shotgun, loads the special launching round, and fires the shotgun. The loader controls the grenades and special launching rounds. One suggested firing sequence is as follows:

1. The firer and loader assume launching positions.
2. The loader loads a grenade into the launching cup and pulls the pin.
3. The loader hands the special launching round to the firer.
4. The firer loads the special launching round into the shotgun.
5. The loader acts as fire-control officer, advising the firer on direction and elevation.
6. When the loader indicates that the weapon is aimed properly and that it is safe to fire, the firer presses the shotgun's trigger, thus firing the weapon and launching the grenade.

Launching cup being attached to shotgun barrel. (Al Pereira photo.)

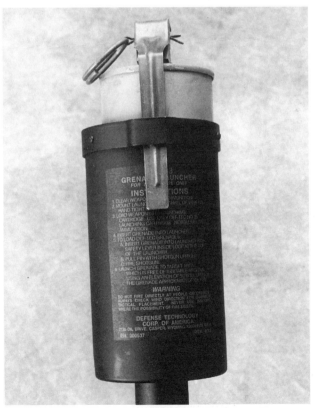

CAW grenade placed in cup. When the pin is removed, the lever remains fixed in the cup channel. When fired, the lever is released and the fuze initiated.

Various shotgun launching cups and launching cartridge.

Close-up of Defense Technology No. 35 Launching Cartridge.

One-Handed Single-Shot Launchers

Illustration 14-4. (Courtesy of DEF-TEC.)

Single-shot, breech-loaded launchers such as DEF-TEC's No. 1314 Gas Pistol illustrated here, are intended primarily for close-quarter tactical situations. Ideal for corrections use, the pistol configuration allows greater ease of movement while still being able to deliver the full range of 37/38mm munitions (including full-size, 8-inch rounds)[4]. The hammer can be manually cocked for single-action fire, or the weapon can be fired from the hammer-forward position using the longer S&W double-action trigger press.

CAW grenade being launched from shotgun equipped with a launching cup. **NOTE:** Hearing and eye protection should be used.

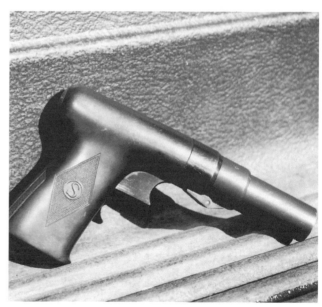

A blast from the past—this small, single-shot gas pistol was produced by the Lake Erie Chemical Company.

Shoulder-Fired Single-Shot Gas Launchers

Sgt. Marty Driggs, MSP armorer, with the single-action TRU-FLITE Super Long Range Gas Gun. The hammer is manually cocked before firing. Many older model launchers such as this are still in service throughout the world. Gloves should be used when firing multiple rounds because the barrel gets hot. A foregrip may also be added. Before firing any launcher, make sure that it is clean and mechanically sound.

The shoulder-fired, single-shot, break-open, breech-loading, smooth-bore launcher has been a staple of U.S. law enforcement since the 1920s. Many of the older models, such as Federal Laboratories' 1.5-caliber Federal Gas Riot Gun and the Lake Erie Chemical Company's TRU-FLITE Super Long Range Gas Gun, are still in service in this country and abroad, mainly because their designs are simple, solid, and effective. This basic design is so dependable, in fact, that contemporary launchers with similar design characteristics are still being produced.

A few common modifications that have been made to the old warhorse include the addition of adjustable foregrips, improved sighting apertures, slings, and high-strength polymer stocks.

With a little practice, you can achieve a reliable degree of accuracy when firing standard munitions through these launchers. For situations requiring precise shot placement, rifled-bore equipped launchers and spin-stabilized munitions are recommended. Regardless of which launcher you use, familiarization with the gear and consistent practice are the keys to safety and accuracy.

Federal Gas Riot Gun

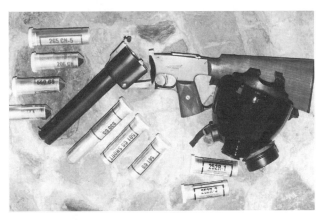

Federal Laboratories' 1.5 Caliber Federal Gas Riot Gun. The hammerless design results in long, double-action trigger pull.

Defense Technology No. 1315 Full-Stock Gas Gun 37/38mm

Defense Technology's No. 1315 Full Stock Gas Gun 37/38 mm is an extremely well-made single-shot launcher. The smooth S&W receiver allows you the option of firing the weapon in double- or single-action modes. The gun has a barrel length of 14 inches (35.6 centimeters), a nonreflective black finish, and a polymer stock and foregrip. It weighs 6.75 pounds (3.06 kilograms) and is also available with a pistol grip stock.[5]

Heckler & Koch 40mm Granatpistole

Heckler & Koch 40mm Granatpistole (Grenade Pistol). Shown here with stock fully retracted. The hammer is cocked, safety is on. Barrel length is 356 mm (14 inches).

The Heckler & Koch 40mm Granatpistole (Grenade Pistol)[6] is extremely well crafted, possessing clean lines and an efficient operating system. It also has the unmistakable look and feel of an HK product. This single-shot, break-open, breech-loading weapon can be fired with the three-position retractable butt-stock at either of two latched-open positions as well as fully retracted. The barrel-tilt locking lever is located directly in front of the manually cocked hammer. To load the weapon, pull the locking lever back, tilt the barrel forward and insert the round. The weapon is also equipped with an ambidextrous safety lever.

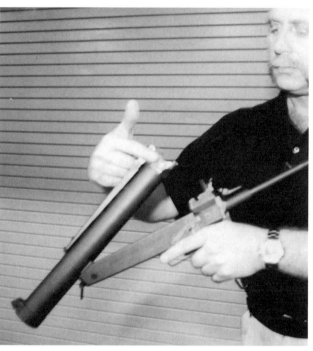

HK's Fred Yates performs safety check with 40mm Grenade Pistol. With safety on and muzzle pointed in safe direction, pull back the locking lever and tilt the barrel forward. Then check visually and physically to ensure that the weapon is safe and clear. This launcher is equipped with an ambidextrous safety lever.

The HK 40 mm Grenade Pistol, right-side view. Shown here with stock fully extended, folding rear sight in elevated position, fixed front sight. This weapon is specially designed to provide a low firing report and recoil as well as superb accuracy. The specimen the author handled was very well balanced: weight, approximately 3 kilograms (8 pounds); length with stock retracted, 517mm (20 inches); length with stock fully extended, 737mm (29 inches).

A closer look at the rear sights. The fixed sight base allows for quick shooting at the 50-to-100-meter range. The folding leaf sight is calibrated for firing at ranges of 150, 200, 250, 300, and 350 meters.

Sage International, Ltd. 37mm SL-1

Sage International, Ltd.'s 37mm SL-1 single-shot gas launcher shown below their modified L-6 version multilauncher. The barrel-release latch is visible on the side of the receiver. The overall length of the SL-1 is 28 inches (70mm); its weight, 6.5 pounds (2.9 kilograms).

Modifications to the single-shot design are apparent in the highly customized 37mm SL-1 produced by Sage International, Ltd. The bore is rifled, an adjustable, tubular cheek rest has been added, and it's been fitted with a redesigned, pistol-grip stock and vertical fore-grip making it very comfortable to mount and fire.

When adjusted for the individual operator, modifications such as these provide for better control of the weapon. This often results in increased operator confidence and accurate placement of munitions.

The SL-1 also features a calibrated sighting system consisting of a flip-up rear sight aperture and flip-up front ballistic ladder that is calibrated from 20 to 100 meters (21.8 to 109.3 yards) in 20-meter increments.

The SL-1 shown here has been designed to fire a special line of ammunition that includes blunt trauma-inducing batons, combination baton/chemical rounds, nonpyrotechnic barricade-penetrating projectiles and pyrotechnic chemical agents. Engravable nylon rotating bands incorporated into the munitions spin-stabilize the projectiles in flight for increased accuracy. The SL-1 weighs in at 6.5 pounds (2.9 kilograms).

LAUNCHING CHEMICAL AGENT WEAPON HAND GRENADES WITH THE GAS LAUNCHER

Single-shot gas launchers may also be equipped with CAW hand grenade launching cups. Procedures are similar to those required when using the shotgun as a chemical hand grenade launcher, although there are some differences in loading methods. After the cup is securely mounted to the end of the launcher's barrel, properly insert the gas grenade and remove the safety pin. Then carefully load the special launching cartridge, close the breech, determine windage and elevation, and you are ready to fire.

Mike Damery launches a CAW grenade from a 37mm gas launcher equipped with a launching cup.

MULTISHOT LAUNCHERS

Several companies in different parts of the world are currently producing multilaunchers. Though they all share similar design characteristics (such as the large, revolver-type, spring-motor-driven magazine and vertical foregrip), the South African-produced Milkor MGL 40mm Multiple Grenade Launcher stands out, thanks to the incorporation of an Armson OEG sight. The Armson sight, which allows the weapon to be aimed while the operator keeps both eyes opened, provides a particular advantage, especially when using the launcher to deliver less-lethal riot control impact munitions.

Model L6, 37mm Multi-Launcher. This multilauncher is lightweight and easy to load and unload. It will accept all standard 37/38mm rounds up to 8 inches (20 centimeters) in overall length. It weighs 3.85 kilograms (10.3 pounds) when empty. The large cylinder must be wound up before firing, much like the Thompson submachine gun drum magazine.

Multishot launchers are extremely well suited for large-scale, riot control applications. Capable of being fired from the hip or the shoulder, they are produced in both fixed- and folding-stock configurations. Though heavier and bulkier than single-shot launchers, the multilauncher's ability to deliver between five to six rounds (depending on model) of 37/38 or 40mm chemical agent munitions in under four seconds has endeared them to operators involved in large-scale civil disturbances.

These launchers also allow for different types of munitions to be loaded and delivered in each application. This feature is especially suitable for delivering a combination of chemical agents and smoke into an objective area.

A close-up of the Model L-6's lower receiver. The safety can be seen here just above the trigger housing.

Regardless of its configuration, loading the breakfront L6 is easy . . .

. . . firing is fast and painless . . .

. . . as is unloading, when the weapon is clean. (Carbon fouling may cause expended cases to stick a bit.)

While they may appear "high-tech," the idea for a multiround gas launcher is not exactly new. In this 1937 photo, an inventor demonstrates an early version of his own design. (Photo from private collection, used with permission.)

OTHER LAUNCHERS

The launchers illustrated in this chapter represent the more common types available today. Numerous models and variations of these models have been produced throughout the world for more than 70 years, though the basic functioning hasn't truly changed.

A few notable launcher variations that exist include the Belgian-made Lansta Pneumatic Grenade Launcher produced by Browning SA. This 48mm low-velocity gun uses compressed air to launch inexpensive grenades up to 150 meters (164 yards). The Lansta can be mounted to a vehicle or used on the ground when fitted to a special carriage.

The French company Alsetex also produces a unique launching system specifically designed for police riot control needs. This launcher is made up of a large diameter tube with a handle mounted at the tail end. Both short- and long-handled versions are available, as is an adaptor that allows the launcher to be mounted to the roof of an armored vehicle. The weapon is designed to launch the company's own 56mm riot control grenades out to ranges of 200 meters (219 yards) but can also be fitted with a subcaliber adaptor that allows it to be used to launch 40mm riot control grenades.

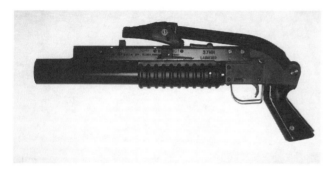

Some commercially made launchers are designed to accept only 37mm munitions of a certain type or length. The Cobray Launcher shown here is one such product. Based on the military M79 "Thumper" 40mm grenade launcher, this launcher's uses are severely limited as far as law enforcement or military applications are concerned.

SAFE CARRY SKILLS

Regardless of the ammunition being used, it must be remembered that all of the launchers shown in this section are capable of firing projectiles of substantial size and weight at velocities that make them extremely dangerous. Though some of the projectiles may be referred to as "less lethal," all are capable of inflicting serious injury or death if fired directly at vital areas at close range.

This fact makes it imperative that we understand what safe carry skills are and that we know how to employ them.

Carry Positions

When carrying any weapon capable of discharging a projectile, you must maintain complete control of it—and yourself. The muzzle of the weapon must never be allowed to cover (point at) anything you do not intend to destroy. Fingers are kept off the trigger and outside of the trigger guard until the decision is made to fire.

involuntary, sympathetic contraction of muscles (in this case the muscles of our arms and hands) that occurs when certain conditions or "stimuli" are present.

As the brain sends signals to the muscles, these signals travel along the nerve fibers of our nervous system and through our spinal cords. Under great pressure or stress, these signals can be sent to either or both of our arms *without conscious determination* on our part.

In most cases the phenomena is initiated when an individual operating under stress is exposed to one of three types of stimuli:

1. *Startle effect.* This effect can be initiated when you are holding a weapon in your hands and an unexpected loud noise occurs, or someone or something suddenly appears, startling you and causing you to flinch involuntarily.
2. *Postural disturbance (loss of balance).* The contraction is initiated if, as you trip or fall, there is anything in or near your hands. The automatic hand-clutching experienced can easily cause any kind of firearm to be discharged.
3. *Exertion of maximum force.* This occurs when you exert great force with one of your hands while holding a weapon in the other. The signals traveling through the spinal cord to one of your arms are also involuntarily sent to the other arm. If you are performing a strenuous clenching action with one hand (such as pulling on a door handle, wrestling with a subject, or even holding back a canine by its leash) while holding a firearm in your other hand, it is possible to experience the involuntary contraction.

If a weapon is being held when any of these stimuli occur and the trigger finger is either on the trigger or finds its way there, it is highly likely that an unintentional discharge may result. It makes no difference whether the weapon is in double-action mode with the hammer down, or single-action mode with the hammer cocked; the contraction experienced as a result of this phenomenon will be intense enough to cause the weapon to fire.

In this series Tim Donnelly demonstrates three acceptable launcher carry positions: low port (top right, previous page), high port (bottom right, previous page), and hip carry (above). Note that his finger is always off trigger and outside of trigger guard until prepared to fire. The "best" carry position depends on factors such as the environment and situation.

UNINTENTIONAL DISCHARGE—A UNIVERSAL DANGER

The possibility of unintentional discharge must be considered when handling any type of trigger-fired weapon. This is especially true when the weapon in question is being carried and/or employed during stressful situations. To reduce the chances of this occurring, we must first understand the difference between an unintentional "involuntary" discharge and an unintentional "accidental" discharge.

Unintentional Involuntary Discharge

A true unintentional involuntary discharge is caused by a phenomena referred to as "sympathetic muscular contraction." It is (as its name implies) an

Unintentional Accidental Discharge

An unintentional "involuntary" discharge can be

attributed to a physiological phenomenon, but an unintentional "accidental" discharge cannot. The sad fact of the matter is that an "accidental" discharge with any weapon usually occurs as the result of a careless, stupid, or negligent act on the part of the operator.

Many lives have been lost or drastically altered as a result of these careless, stupid, and negligent acts, and many careers have been ruined.

So slow down, stay alert, and think twice before doing anything when handling any type of firearm. And be sure that you *and those around you* are familiar with (at a minimum) the basic rules of safe firearms handling.

BASIC RULES OF SAFE FIREARMS HANDLING

1. Treat all firearms as though they were loaded— ALL OF THE TIME!
2. NEVER allow the muzzle of any firearm to point at anything you are not willing to destroy.
3. Keep your finger outside the trigger guard until you are on target and have decided to fire.
4. Be sure of your target and what is beyond it.

ADDITIONAL COMMONSENSE RULES FOR SAFE FIREARMS HANDLING

1. NEVER handle any firearm if you are not sure of what you are doing.
2. Know how to perform a safe condition check: To perform: Immediately upon picking up a firearm you are familiar with, (a) keep it pointed in a safe direction, (b) remove the magazine or feed tube if it is so equipped, (c) open the action and remove any ammunition, (d) visually and physically check the breech or cylinders to confirm that the firearm is unloaded, and (e) check a second time.
3. Except in an emergency, never give any firearm to, or take one from, ANYONE, unless the action is open so you can see that it is clear of ammunition (as described in #2 above).
4. Unload firearms when not in use.
5. Store weapons and ammunition separately.
6. Use only quality ammunition.

7. Never perform any type of strenuous action (such as grabbing or pulling) with one hand while holding a loaded firearm in the other if you can avoid it. These actions may cause you to fire the weapon unintentionally.
8. Never permit the muzzle of a firearm to touch the ground.
9. Avoid any kind of drug or alcoholic beverages when handling firearms.

SAFE DIRECTION: A DEFINITION

A safe direction is defined as any direction the weapon may be pointed where a discharge would result in no loss of life, no injury, and limited property damage.

CARE AND MAINTENANCE

Gas launchers must be maintained to ensure years of problem-free service. Many of them will sit for long periods between uses, so they must be regularly inspected for rust or dust buildup.

Launchers should always be inspected before firing, and all precautions used in firearms handling must be observed.

After use, the launchers must be cleaned, lubricated, and stored properly. When cleaning, pay special attention to the barrel and breech face, especially if the propellants used in the munitions contain salt.

Since salt is a corrosive that will eat away at the very pores of the metal, it must be completely removed. The best way to remove salt is by cleaning these areas thoroughly with water—another enemy of metal. What this means is that after the bore and other surfaces have been cleaned with water, all the water must then be removed. The surfaces can then be cleaned with solvent to remove the carbon deposits that will resist the water.

After thoroughly cleaning out the carbon, a light coat of oil or gun grease should be applied to all metal surfaces.

A high-viscosity gun grease, such as RIG, is

preferable to oil because it lasts longer and maintains its consistency. I've also used MILITEC-1, a synthetic metal conditioner for a number of years with great success.

Storage

Ideally, gas launchers should be stored in a clean, dry environment. Launchers that are routinely stored in vehicles must be checked more frequently than those maintained in humidity-controlled arms rooms because trunks often leak or absorb moisture from damp items that may find their way inside.

One of the worst feelings in the world is to pull your issued launcher or shotgun out of its carrying case only to discover that it is covered with surface rust.

Besides the fact that you are personally responsible for the gear in your possession, the fact is that long-term storage neglect can render the equipment useless. Should you find yourself holding an inoperative weapon at a scene during a crucial moment, you may find yourself being considered useless as well.

NOTES

1. People's Republic of China.
2. The less lethal kinetic energy grenade launches a rubber ball projectile.
3. Dye-marker grenades are designed to stain riot participants so they may be identified after order is restored.
4. Because the gas pistol's barrel itself is only 8 inches long, accuracy with the full-sized rounds is not as reliable as with shorter rounds.
5. When fitted with the pistol grip stock, the weapon is designated as the No. 1316 Tactical Gas Gun.
6. Though designated a grenade "pistol" by the manufacturer, the HK 40mm launcher is included in this section because it can be shoulder fired when the integral stock is extended.

Weapon Retention

When engaging in any activity that requires you to use handheld weapons against others, *especially* while in close proximity to them, you must always remember that any one of them may decide that your weapon would serve much better if it was *his* or *her* weapon. If this occurs, you may well find yourself fighting for control of both the weapon and the situation.

The bottom line is that your weapon remains yours only if you know how to hang on to it. Better yet, if you can discipline yourself to stay alert and always present a professional appearance, you may be able to discourage someone from even attempting to make a move against you.

ASR RETENTION

If you decide to employ an ASR during an encounter, you must first get it into your hand. This requires you to remove it from the holster, pouch, or pocket you carry it in. Once in your hand, the ASR, just like a firearm, is under your control only so long as you can keep it in your possession.

Although a good deal of firearms training is devoted to impressing this fact upon rookie and veteran officers alike, the idea of someone wrestling for control of your ASR is rarely touched upon.

This is a mistake.

First, common sense indicates that if there are people out there who will risk being shot while attempting to take a firearm from a trained police officer, then chances are great that there are probably more people who would consider a grab for your ASR worth the risk of being sprayed.

Second, should someone attempt to take the ASR from your hand, it is possible that you may fixate upon retaining that ASR and miss other danger cues or threat indicators, such as the offender's changing his focus to your holstered pistol.[1]

Should this occur, you may find yourself unable to process all the danger cues and information properly and therefore responding improperly or, worse, entering a state of hypervigilance, rendering you unable to respond at all.

To combat this possibility, you must first recognize that it exists. You must then consider what options are available to you should it happen. Finally, you must rehearse your response options in a controlled environment so you will know what you can realistically do. Your agency or department should address these things during training. But if it does not, you can do it on your own, perhaps practicing some techniques with a fellow survival-minded colleague.

It's not rocket science. Sometimes the simplest consideration or basic practice will pay off tenfold if you find yourself face-to-face with the unthinkable.

While defending your ASR from an assailant, don't forget about the possibility of his going for your other weapon: your handgun.

GRENADE RETENTION

Even though CAW grenades are usually employed from a distance, the subject of retention cannot be overlooked, especially when you are operating in heavily congested environments such as a riot or other disturbance. Under these conditions someone may try to take or knock the grenade from your hand. You must be sure to have a good grip on the grenade while moving or otherwise preparing to deliver it to the objective area. Dropping *any* spoon-equipped grenade after the pin has been removed but before you're ready can be problematic. (Should this occur, you might have time to pick up the grenade and immediately deliver it to the objective area before it goes off, depending on the fuse delay and type of device.) Suddenly finding yourself facing an immediate deadly threat is another consideration that must be taken into account, especially since most people deliver the grenade while holding it in their dominant hand. What do you do if you need to

switch to the pistol? Do you simply drop the grenade and run for cover while accessing your firearm? Throw the grenade at the assailant and then run for cover while reaching for your firearm? Switch the grenade from one hand to the other while running for cover and reaching for your firearm? All of these are viable options. If you think about them, maybe practice them in training once or twice, you'll discover which works best for you. If you don't think about it or practice, chances are you'll stand there with your mouth open and the grenade firmly clasped in your hand and let someone else take control of your destiny.

And that is unacceptable.

LONG-GUN RETENTION

Any competent police handgun-training program will address the issue of weapon retention, but the idea of long-gun retention is often overlooked.

Regardless of whether the weapon is a rifle,

shotgun, submachine gun, or gas launcher, consideration must be given to the possibility that someone may try to take it, as well as to responses to prevent that person from succeeding. While some strides have been made in this area in regard to close-quarter submachine gun work, retention considerations for other long guns often remain neglected.

I believe that part of the reason this subject is overlooked is the nature of our business when we use long guns. That usually (unless performing as part of an entry team) means that we are operating at greater distances from the adversary than we normally do. In most cases we will be better prepared physically and psychologically for the engagement: the very presence of the long gun in U.S. police work indicates that there has been some warning given to allow the officer time to get it. And in many cases, I believe, officers armed with a long gun tend to feel less vulnerable simply because of the confidence-inspiring presence and intimidating appearance of the weapon.

It is critical to remember, though, that there are

If you are attacked by someone attempting to disarm you while you are equipped with any long gun, you must react quickly and fluidly. In this example, the gas gun itself is used to strike the offender in the most efficient manner available to the officer, based on position and configuration of the weapon and the offender's attack posture.

. . . the officer simultaneously switches his grip on the thumb-hole-configured stock to give him more purchase and better control of the weapon.

First, the officer steps back, pulling the offender off balance and twisting the weapon out of the offender's hands as . . .

The weapon is then fluidly rotated, and a modified horizontal buttstroke is delivered, allowing the officer to maintain possession of the weapon and create distance between himself and the offender. The officer then prepares to follow up in case the combination of blows does not effectively end the attack.

many individuals out there who may not hesitate to take on an officer, no matter how well trained. Many people who have served as members of their country's military forces have been trained in methods that can be used to disarm and disable a heavily armed opponent. People trained in various martial arts may also not hesitate to attack or disarm you. Others may also not be deterred by the presence of any type of weapon, depending on the situation, your actions/reactions, and their own mental state.

So again, stay alert, consider your options before you need to exercise them, and take nothing for granted.

Fogger officer with security element.

FOGGER

Much like the long gun, the fogger has an imposing appearance. The heavy volume of smoke or irritant it produces also tends to make people want to be anywhere but where it is. However, one consideration must be that the fogger may be used fairly closely to the subjects it is deployed to disperse.

Since recent bouts of civil unrest in the United States and abroad indicate that chemical agents may not be as psychologically effective as they once were, this closeness could pose both a significant retention and personal safety problem for the officer charged with operating the sometimes cumbersome foggers. This seeming loss of the ability to intimidate may be due greatly to the information available to people via the Internet. There are numerous Web sites that not only teach how the police and chemical agent weapons are employed, but also how to combat them during violent protests and riots.

Items such as gas masks, chemical agent weapons, and bullet-resistant vests are also freely available to practically anyone who desires them, all with the click of a button.

The violent activities accompanying the World Trade Organization riots in Seattle and elsewhere, for example, also indicate that protesters are becoming more organized and sophisticated and less likely to run away at the first signs of chemical agent weapon or smoke clouds and helmeted response teams.

The danger levels for CAW-equipped officers will also increase should any demonstrator organizations adopt a more violent approach or be infiltrated by hard-core extremists who take advantage of peaceful protests to engage in violent aggressive behavior—as has happened.

For these reasons, it is recommended that officers equipped with any less-lethal equipment be protected by dedicated security officers. Both the CAW officers and security elements officers should be allowed to work together during training and operations to iron out any logistical or communications problems. In this way, both the less-lethal mission and the officers assigned to carry it out will be protected.

NOTE

1. There have also been reports from various correctional institutions that some convicts have been observed practicing "feints," such as grabbing an officer's holstered pistol and then immediately removing another weapon (such as an exposed-carry knife) from the officer's person and using it against the officer.

Tactical Employment of Chemical Agent Weapons Outdoors: An Overview

This chapter is intended to provide a brief overview of the subject. Prior to employing any tactics, chemical agents, or riot control techniques, proper instruction from certified and experienced trainers must be sought. Other texts specifically dealing with riot control materiel and techniques are suggested in the bibliography.

SITUATION EVALUATION AND TACTICAL ANALYSIS

CAWs are used in outdoor crowd control situations to induce people to move. If CAWs are used improperly, the results of this movement can be disastrous. Therefore, before any CAWs are employed to move crowds of people, it is imperative that there be a plan to ensure a clearly defined purpose for CAW use.[1]

Specifically, CAWs are used to break up unlawful gatherings of people and induce them to move to a planned escape route. In this way, the unruly group can be dispersed, individual identified agitators can be taken into custody, and order can be restored.

To accomplish these objectives in a reasonable and professional manner, the concentration levels of the applied agents must be controlled, and their deployment must be planned and strategically executed. Before this is possible, the situation needs to be thoroughly evaluated and recommendations made to the designated incident commander.

Consideration must also be given to the composition, intent, and activities of the crowd to

Officer in basic riot gear, with protective mask in carrier on hip. All officers should have chemical agent protective masks available whenever CAWs are to be used. If the masks are not worn at the time the agents are deployed, officers must have them on their person in an accessible location so they can be donned immediately.

CAWs should be experimented with during training in different weather, climate, wind, and temperature conditions. In this way, officers can gain experience and see how the agents and delivery devices are affected by these differences.

determine whether the use of CAWs is warranted and, if it is, to further determine the safest, most efficient way possible to employ them.

When evaluating the situation, be sure to include any information regarding buildings or other structures in the area. Rioters may get inside these structures or up on rooftops. Schools, hospitals, or nursing homes may be directly behind the OA, or even inside it. The following factors must be anticipated and prepared for: terrain, wind, temperature, humidity, thermal turbulence, and precipitation.

Terrain

The terrain is critical to deciding how, when, and where to employ CAWs. Devices tend to bounce and move more on such hard surfaces as pavement and concrete than they do on soft or grassy areas. The presence of hills, water, trees, buildings, bystanders, or traffic must also be considered because smoke-borne agents and aerosols will travel differently, depending on the type and configuration of the terrain and its features.

Clearly defined escape routes that are available and obvious to the crowd must also be identified before agents are deployed. This is critical to ensuring that the crowds do not run toward the officers in a panic once exposed to the effects of the CAWs because this may result in unnecessary injuries being incurred by both police and citizens.

Wind

Wind speed and direction are the primary considerations when using munitions outdoors because they determine when, where, how, and *how much* agent should be released.

Ideal wind conditions are defined as 5-to-10-mph winds blowing toward the crowd and the designated escape routes and *away* from the police line (see Illustration 16-1). These conditions should result in rapid coverage of the OA by a sufficient concentration of agent. Faster moving winds will cause the agent to dissipate faster and travel farther—possibly farther than is desired. In cases such as these, the release line (RL) will need to be moved back farther from the OA, and more munitions will be needed.

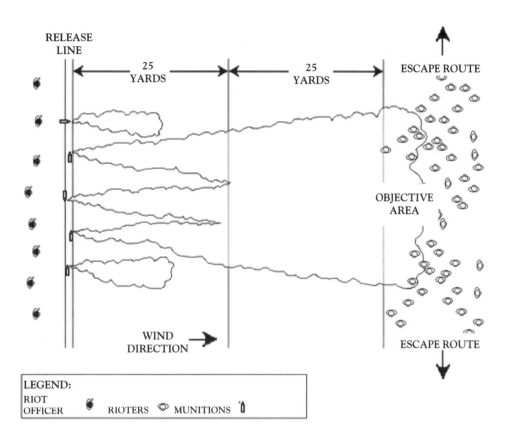

Illustration 16-1. Grenade RL placed 5 yards (4.6 meters) apart will form a continuous cloud approximately 25 yards (23 meters) forward of the release line. If crowd is 150 yards (137 meters) downwind of police line, the release line should be 125 yards (114 meters) in front of officers.

If there is no wind blowing, then the RL will need to be moved closer to the OA and the crowd. Thermal foggers may be used (see Illustration 16-2). CAW grenades/devices may also be bounced or rolled into the crowd. It is imperative that chemical agent burst-type grenades, if used, be deployed so that they burst on the ground, not while airborne. While some muzzle-blast dispersion agents may be fired into the crowd at a low angle and only from recommended distances, *no grenades, projectiles, or any other devices should be fired directly into the crowd because serious death or injury may result.*

If the wind is blowing from behind the crowd and toward the police, then the agents must be released from behind the crowd so it will blow through them and toward the escape routes and the police line (see Illustration 16-3). If the RL is from 50 to 150 yards (45.7 to 137 meters) from the police

line this can be accomplished by launching CAW grenades from the police line so they land behind the crowd—but, again, you must never throw or fire any type of device directly into a crowd unless deadly force is warranted. Once the proper concentration of agent is determined and released, the cloud must be maintained by replacing munitions until the crowd disperses. Burn times for individual types of munitions must be known and used to replace munitions.

Finally, if the wind is blowing from the right or left flank in a crosswind situation, then the agent should be released to the appropriate side so it will be carried by the wind toward the crowd and the escape route or routes (see Illustration 16-4).

Lateral wind currents or speed will also affect the way the agent cloud moves downwind, causing it to blow from side to side.

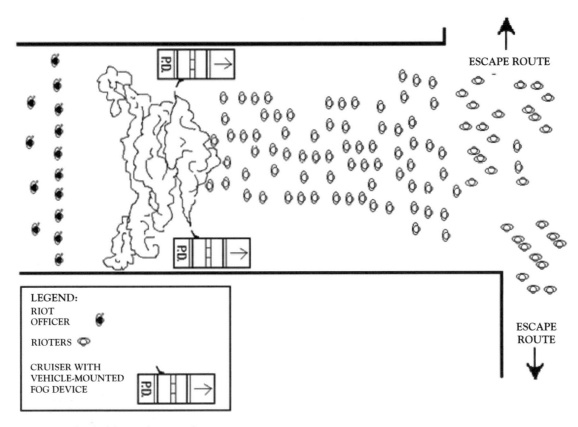

Illustration 16-2. Thermal foggers being used.

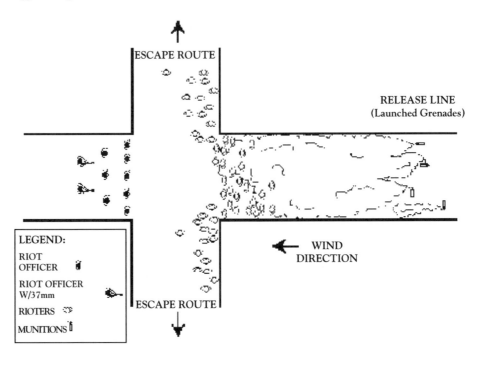

Illustration 16-3. If the RL is 50 to 150 yards (45.7 to 137 meters) from the police line, CAW grenades can be launched into RL

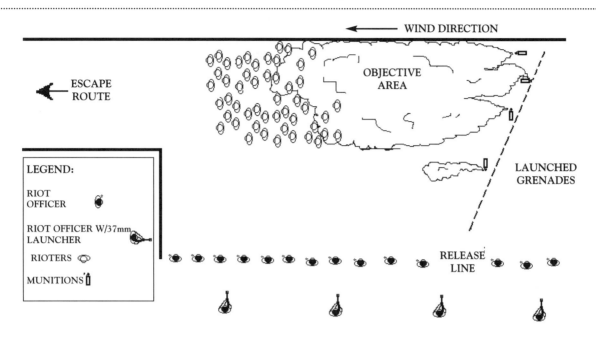

Illustration 16-4. Depending on the distance from the police line to the RL, CAW grenades can be thrown or launched into RL position.

Estimating Wind Speed and Direction

One method that may be used to estimate wind speed is illustrated in the photos below. Any flags located in the immediate OA can be used. You simply determine the approximate angle between the flag and the pole (in degrees) and then divide that number by four. The result gives you a rough

Using flags to estimate wind speed and direction. The degree of angle and wind velocity will be approximate values. (A protractor can be used to determine the angles of the flag in these photos.)

estimate of the approximate wind speed in miles per hour. A variation of this technique that doesn't require a flag can also be used to determine approximate wind speed. You simply hold a small piece of paper, grass, cloth or other light material at shoulder height while standing. Then open your hand and let the object fall from your hand. Without moving from your location, keep your arm straight, point directly at the object, and then divide the angle between your arm and body by four, as you did with the flag technique.

Wind direction can also be estimated by observing the direction in which flags or other objects are blown. However, since many factors can affect wind currents and direction, the most reliable way to estimate the best place to release the agents and their most probable direction of travel is by releasing smoke before the chemical agents.

Using Smoke

If smoke is used to determine prevailing wind currents, it must be remembered that the mere appearance of smoke—particularly colored smoke, according to the late Col. Rex Applegate—may induce a psychological reaction up to and including panic from the crowd.

If the prevailing wind is favorable, smoke can be used to split a large crowd into two divergent groups. Smoke can be released using grenades, candles, or

Smoke pots can be emplaced by hand or dropped off the backs of vehicles or from helicopters. They generally burn from 5 to 15 minutes and produce extremely large volumes of obscuring smoke. The example shown here is an older smoke pot.

larger smoke pots. This tactic, combined with the obscuring effect of the smoke screen, may cause rioters to become confused and lose sight of one another. Once this occurs, the ability of significant numbers of rioters to join in a controlled, concerted effort is severely hampered.

Smoke can also be combined with CAWs for certain applications. The recommended method is to release the smoke and CAWs into the OA while officers maintain a perimeter around the OA. Police officers should not enter the OA while visibility is diminished. Rather, they should maintain perimeter positions and take rioters into custody as they emerge from the OA.

Temperature

Temperature affects the lateral spread, vertical height, and speed of dissipation of the smoke or CAW cloud, especially if the temperature is markedly different at different elevations.

The different temperature levels, referred to as gradients, are measured by subtracting the air temperature 1 foot (30 centimeters) above the ground surface from the air temperature 6 feet (1.8 meters) above the ground surface.

A device for measuring gradients that can be simply produced is shown in Illustration 16-5.

There are three types of temperature gradients: inversion (or stable), neutral, and lapse (or unstable).

Inversion, or Stable

A stable condition exists when the air temperature increases with an increase in altitude. Stable air currents are produced when this happens, often resulting in the smoke staying close to the ground. The streams of smoke also tend to spread and diffuse slowly.

Neutral

Neutral gradients exist when there is little difference in air temperature at different altitudes. Neutral temperature conditions provide the best conditions for the use of smoke because they allow the smoke to travel downwind in a steady direction while spreading and rising fairly quickly.

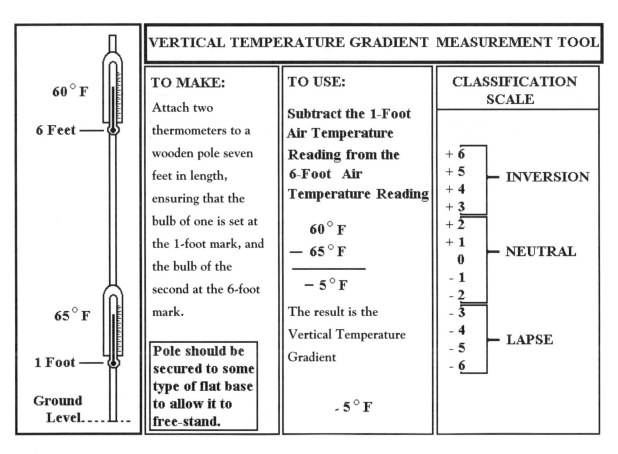

VERTICAL TEMPERATURE GRADIENT MEASUREMENT TOOL

	TO MAKE:	TO USE:	CLASSIFICATION SCALE
60°F 6 Feet ⎯⦿ 65°F 1 Foot ⎯⦿ Ground Level ⎯⎯	Attach two thermometers to a wooden pole seven feet in length, ensuring that the bulb of one is set at the 1-foot mark, and the bulb of the second at the 6-foot mark. **Pole should be secured to some type of flat base to allow it to free-stand.**	Subtract the 1-Foot Air Temperature Reading from the 6-Foot Air Temperature Reading 60°F — 65°F ——— — 5°F The result is the Vertical Temperature Gradient - 5°F	+ 6 ⎤ + 5 ⎥ — INVERSION + 4 ⎥ + 3 ⎦ + 2 ⎤ + 1 ⎥ — NEUTRAL 0 ⎥ - 1 ⎥ - 2 ⎦ - 3 ⎤ - 4 ⎥ — LAPSE - 5 ⎥ - 6 ⎦

Illustration 16-5. Vertical temperature gradient measuring device.

Lapse. Unstable air currents are produced, resulting in the smoke rising abruptly from the source and quickly dissipating.

Lapse, or Unstable

An unstable condition exists when the air temperature decreases with an increase in altitude. Unstable air currents are produced when this happens, often resulting in the smoke rising abruptly from the source and dissipating quickly.

Mechanical Turbulence

Mechanical turbulence occurs when irregular movements and flows of air are created as a result of wind currents contacting surface obstacles. When this happens, aerosol and smoke-borne agents may flow unpredictably into cellars, indentations on the ground, and other such places. (See Illustration 16-6.)

Humidity

Humidity (the amount of moisture in the air) tends to hold agents closer to the ground. High

145

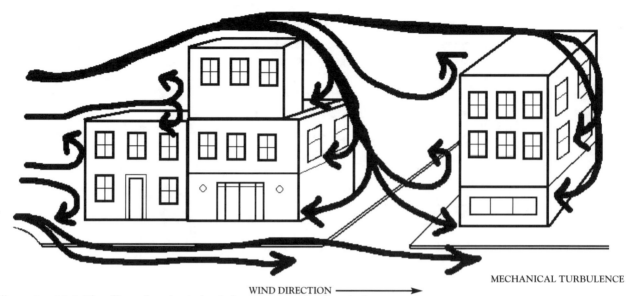

WIND DIRECTION ⟶

MECHANICAL TURBULENCE

Illustration 16-6. The effects of mechanical turbulence on aerosol and smoke-borne agents.

THERMAL TURBULENCE

Illustration 16-7. The effects of thermal turbulence on aerosol and smoke-borne agents.

humidity also tends to increase the agent's burning effect on exposed skin.

Thermal Turbulence

Thermal turbulence is caused when descending cooler air displaces rising warm air, and this creates irregular flow patterns. When this occurs, aerosol and smoke-borne agents may travel unpredictably upward along the sides of structures and enter such access points as windows, doors, and crevices. (See Illustration 16-7.)

Precipitation

Rain or snow will force agent particles to the ground, severely limiting their effectiveness.

OUTDOOR INCIDENT PREPARATION AND EMPLOYMENT GUIDELINES

1. **Preparation of Riot Control Personnel**
   a. Personnel should be prepared as far in advance as possible.
   b. Proper equipment should be provided, checked, and used to train.
   c. Movements and formations should be coordinated and rehearsed.
   d. Intelligence about the nature and scope of possible riot/disturbance activities should be developed.
   e. Comprehensive briefings should be organized and conducted for all involved personnel. Officers should be advised of potential dangers, procedures, and contingency plans.
   Don't leave your people uninformed!

2. **Incident Notification Checklist**
   a. Command officers
   b. Chemical munitions officers
   c. SWAT team officers
   d. Mounted unit officers
   e. K-9 officers
   f. Local police agencies
   g. State/county police agencies

3. **Establishment of Inner Perimeter at Scene**
   a. Ensure that you have enough officers for the job.
   b. Ensure that you have the right equipment for the job.
   c. Ensure that you have enough equipment for the job.
   d. Ensure that leaders are competent.
   e. Ensure that leaders are known/identified to personnel.

4. **Establishment of Outer Perimeter around Scene**
   a. Reserve manpower staging area must be designated.
   b. Sheltered break area/latrine areas must be designated.
   c. Arrangements for food and water for extended incidents must be made.
   d. Arrangements for relief of line officers must be made.
   e. Arrangements must be made for changing CAW-contaminated uniforms.
   f. Arrangements for cleaning CAW-contaminated uniforms must be made, especially if incident is anticipated to go on for an extended period.

5. **Evacuation of Injured Victims**

6. **Evacuation of Bystanders**

7. **Establishment of Central CP and Chain of Command**

8. **Request for Ambulance, Rescue, or Fire Equipment**

9. **Authorization for News Media Access and News Media Policy**

10. **Authorization for Use of Force and Chemical Agent Weapons**

11. **Establishment of Interaction between CP and Riot Control Elements and Responsibilities of Each**

12. **Establishment of Plan to Move Rapidly into OA after Crowd Is Dispersed in Order to Remove Lingering Groups and Prevent Crowd from Regrouping**

13. **Establishment of Contingency Plans**

POST-INCIDENT GUIDELINES

These are provided for reference only.

1. Roll Call and Wellness Check of All Personnel

2. Area Check to Ensure No One Injured or Hiding Is Left Behind

3. Decontamination of Personnel and Equipment

4. Collection of All Expended Munitions

5. Inventory of All Munitions Used and Not Used

6. Written Report Detailing Order of Events and Actions Required/Taken

7. Report on Effectiveness of Agents and Delivery Systems Used

8. Medical Review by a Qualified Physician of All Persons Exposed to Any CAW Recommended

NOTES

1. CAWs should also be experimented with during training in different weather, climate, wind, and temperature conditions, so officers can gain experience and see how they act and react.
2. Colonel Applegate noted that this panic would be induced partly because people believed that the colored smoke had dye qualities that would indelibly mark them for later identification. This belief does have some basis in fact: stains may result on clothing, skin, and objects if conditions are right. The Colonel also believed that colored smoke produced a sort of "voodoo effect" on people and that part of the reason they would react so strongly was out of a deep-seated fear that the smoke would be sickening or poisonous.

Tactical Employment of Chemical Agent Weapons Indoors: An Overview

This section provides a brief overview of the subject. Before employing any tactics, chemical agents, or techniques in the field, officers should seek proper instruction from certified and experienced trainers.

SITUATION EVALUATION AND TACTICAL ANALYSIS

The decision to use any CAW indoors depends on many factors. The type of CAW required for the specific situation at hand will determine the level of authority needed to sanction its use, as well as affect the way it is employed.

The justified use of ASRs, for example, will in most cases be determined by the individual officer on scene and applied as required. The use of any other type of CAW, however, generally needs to be approved by a higher level of authority and be employed by personnel specially trained to do so.

NOTE: *Excluding ASR-type munitions, CAWs should generally be used only as the last tactical option before resorting to deadly force in any given situation.*

Before the use of CAWs, every attempt should be made to persuade any barricaded subject to come out of a building or structure.

After a perimeter has been established, you should gather as much pertinent data about the suspects, hostages, and OA as possible.

If possible, contact should be made from a safe location with barricaded suspects. Telephone, loudspeakers, or vocal communications from behind cover can be used. *Officers must always be aware of potential dangers and use cover as appropriate.*

Officers should try to avoid inducing panic in barricaded suspects, especially if there are hostages inside.

Medical personnel, fire apparatus, and an ambulance should be standing by at the scene prior to any CAW use, indoors or outdoors.

PREPARATION

Before using a CAW during a situation involving a barricade or hostage, a whole list of preparations must be made. An inner perimeter must be established around the OA, ensuring the containment of suspects and security against others' trying to enter the OA.

An outer perimeter must be established to bring the situation under greater control. Traffic must be diverted from the area, and injured victims and bystanders must be evacuated. A command post (CP) and chain of command need to be established. Medical personnel, fire apparatus, and an ambulance should be standing by at the scene or close by.

The designated incident commander or higher authority must also give authorization for use of force and chemical agents.

Establishment of interaction and communications between the CP, special operations (or SWAT) unit, and hostage negotiation unit must be made, and the responsibilities of each clearly defined.

A hasty plan, an operational plan, and contingency plans must be formatted and, when possible, rehearsed prior to deployment.

Authorization for news media access and news media policy must be addressed.

Liquid or expulsion munitions are preferred for indoor use. Pyrotechnic munitions other than the "flameless" type that are designed for indoor use should not be introduced into—or under, or on top of—a structure because of the danger of fire.

You must remember that most liquid projectiles and grenades produce invisible vapors, making it hard to tell by visual cues alone whether the agent has been dispersed. In many cases, the suspect's actions will be the most obvious indicator of the success of the CAW application.

CALCULATING FOR SAFETY: LCT$_{50}$ AND ICT$_{50}$

Before introducing CAWs containing CS or CN, lethal concentrations and time (LCt$_{50}$) and the median incapacitating dosages (ICt$_{50}$) of these munitions must be calculated and double-checked (see Appendix A). The decision regarding the type and quantity of munitions to be used should be based on these calculations. That these calculations were used should be documented and included in the final written incident report.

RESPONSIBILITIES

If the suspect ceases to respond and cannot be located during the process of applying a CN or CS CAW, it must be assumed that he is in danger of overexposure. Though this is rare, there have been cases where subjects have died after exposure to a CAW-contaminated environment for extended periods in confined spaces. That is one of the reasons

that tactical entries shortly after the introduction of CAWs are often factored into the operational plan.

Although there is an absolute obligation on the part of the officers introducing the CAW into a building to ensure that anyone inside is not placed in danger from prolonged overexposure, there is also an absolute obligation for the officers to protect themselves and others from potential dangers posed by the suspect(s) who initiated the incident to begin with.

Just as with the suspect who chose to exit from the window on the seventh floor rather than be taken into custody, some determined individuals may decide that the fight will not be over until they are forcibly removed from their barricaded position, which can mean captured, subdued, or killed. Regardless of how irrational that may seem to you or me, we must remember that if someone has decided that he is willing to risk (or in some cases take) his own life for whatever reason, then he probably isn't going to be much concerned about the lives of the people trying to get him to come out.

As for suspects not responding to CAW applications, one possibility is the device's malfunctioning after being introduced into the OA. It is also not unheard of for barricaded suspects to take such countermeasures against chemical agents such as placing blankets or mattresses against windows and doors or putting wet rags over their faces to slow the affects. In a few cases I'm aware of, barricaded suspects were prepared for an extended conflict, even to the point of having laid in extra food and water supplies and equipping themselves with gas masks and weapons. The possibility of booby traps and ambushes inside the barricaded area also exists and cannot be overlooked.

ENDGAME

The introduction of CAWs into a building containing a suspect generally produces one of the following results or a combination thereof.

Suspect Surrenders

If the suspect surrenders, he should be ordered outside and into a standard control position (e.g.,

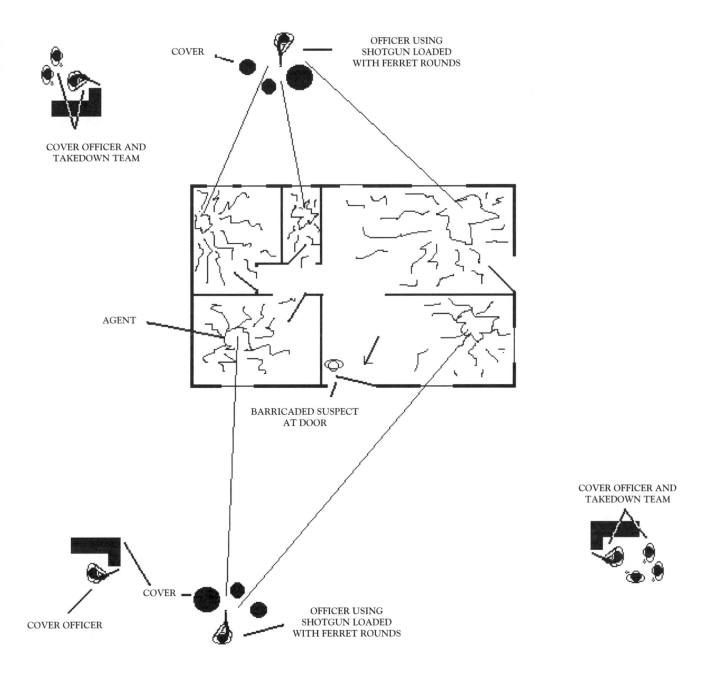

Illustration 17-1. Officers fire nonpyrotechnic, barricade-penetrating ferret rounds into a house while cover officers and takedown teams wait in position. CAWs should be applied methodically if possible, room by room, with officers stopping periodically to encourage the suspect to leave the structure through an available exit. Regardless of how carefully you plan and execute the operation, you cannot anticipate how the suspect will react to the introduction and effects of the CAW. I've witnessed such diverse reactions as suspects stumbling out a door still looking for a fight; burying themselves under piles of dirty clothing; driving their heads through multiple glass windows and window screens to get to fresh air; and, in one memorable incident, tying bed sheets together to form a line and going out the window of a hotel with smooth outside walls—from the seventh floor! Despite the fact that the sheets only reached the fifth floor, the highly motivated suspect dropped to a second-floor roof and then jumped the rest of the way to the ground and took off on foot. (They told me if I stayed on the job long enough, I'd see just about everything. . . .)

kneeling with ankles crossed, hands on top of head; or prone with arms extended, palms up), handcuffed, and searched thoroughly. Medical assistance should then be administered if necessary.

If the suspect indicates that he wishes to surrender but is unable to navigate to the outside due to the effects of the CAW, then an appropriately trained and equipped entry team will have to be used to secure the suspect and remove him from the building. The threat of ambush or booby trapping must be considered.

Suspect Runs from the Building

If the suspect runs from the building in an attempt to attack, escape, or surrender, or as a result of panic, officers should neutralize or intercept him preferably before the suspect passes through the inner perimeter. The decision to use any level of force against the suspect is determined by the suspect's actions (i.e., the suspect armed or otherwise presenting a deadly threat as opposed to running with his empty hands rubbing face and eyes). Real-time communications should be established to alert those on the outer perimeter of the suspect's activities, location, and direction of travel, should he penetrate the inner perimeter. When it happens it usually happens *fast*.

Suspect Remains in the Building

If the suspect remains in the building after CAWs are introduced, he may choose violence as an option. The decision to inflict death or injury on himself or others may be made in a blind moment of panic as the CAW devices come through the windows, or may have been reached days, weeks, or even months earlier. The possibility of violence must be considered and prepared for before CAWs (or entry teams) are introduced into the OA. Long-range tactical marksman/observer teams should be in place. If hostages are present, a highly trained entry team should be standing by. All officers at the scene should find solid cover and ensure that they are not exposed. (One of my teammates possibly saved his own life one afternoon when he decided to move to a better position of cover moments before CAWs

were introduced into an armed suspect's barricaded location. Immediately after the CAWs were deployed, the suspect opened fire with an SKS rifle, and one of the rounds struck the pavement where my teammate had been.)

If a suspect remains in the building after CAWs are introduced and continues to resist verbally or physically, then a reassessment must be made. Considerations that should be taken into account before further action is decided include the following:

- Did the CAW fail?
- Were the CAWs deployed in the right locations inside the structure?
- Were the CAWs deployed in sufficient quantities based on LCt_{50} and ICt_{50} calculations?
- Were such countermeasures as placing blankets or mattresses against windows and doors taken by the suspect?
- Does the suspect possess any chemical agent mask, either commercially produced or improvised?

If the suspect remains in the building but cannot be seen or does not respond to verbal communications, another reassessment must also be made. In addition to the considerations discussed above, the decision to enter the building (and the procedures used in doing so) to locate and secure the suspect must be made based on all the circumstances present in a particular situation.

SITE DECONTAMINATION: AN OVERVIEW

Most people consider the successful removal of a barricaded suspect or suspects as the completion of the operation, but it's actually not—at least not if CAWs have been employed. After a suspect has been attended to and transported from the scene, there are still the matters of CAW munition recovery and decontamination to deal with.

Personnel decontamination methods were discussed in Chapter 11, so this section will provide an overview on site decontamination.

Decontaminating a house, an apartment, or a building that has been saturated with CAWs

INDOOR INCIDENT PREPARATION AND EMPLOYMENT GUIDELINES
These are provided for reference only.

1. **Activation of Special Operations Team Personnel**

2. **Incident Checklist Notification**
   a. Command officers
   b. Hostage negotiator
   c. K-9 officers
   d. Local police agencies
   e. State police agencies

3. **Establishment of Inner Perimeter at Scene**

4. **Establishment of Outer Perimeter around Scene**

5. **Evacuation of Injured Victims**

6. **Evacuation of Bystanders**

7. **Establishment of Central CP and Chain of Command**

8. **Request for Ambulance, Rescue, or Fire Equipment**

9. **Authorization for News Media Access and News Media Policy**

10. **Authorization for Use of Force and Chemical Agents against Barricaded Suspect**

11. **Establishment of Interaction among Command Post, Special Operations Team Personnel, and Hostage Negotiation Personnel and Responsibilities of Each**

12. **Establishment of Hasty Plan and Operational Plan**

13. **Establishment of Contingency Plans**

POST-INCIDENT GUIDELINES
These are provided for reference only.

1. **Roll Call and Wellness Check of All Personnel**

2. **Thorough Building and Area Check to Ensure That No One Injured or Hiding Is Left Behind**

3. **Decontamination of Personnel and Equipment**

4. **Collection of All Expended Munitions**

5. **Inventory of All Munitions Used and Not Used**

6. **Written Report Detailing Order of Events and Actions Required/Taken**

7. **Report on Effectiveness of Agents and Delivery Systems Used**

8. **Medical Review by a Qualified Physician of All Persons Exposed to Any CAW Recommended**

containing CN, CS, or OC can be time consuming and expensive. One option is to use commercial cleaning companies specially trained and equipped to deal with the situation. But anyone involved in the decontamination process has to be equipped with protective clothing and equipment. At the minimum, a respirator with appropriate filters, protective gloves, boots, and clothing will be needed.

After opening all doors and windows and shutting off air-conditioning systems, the first thing to do is retrieve all canisters, casings, or munitions. These articles should be recovered by the involved department's assigned chemical agent weapons officer (CAWO), inventoried, and disposed of in accordance with all applicable regulations and policies.

If CN, CS, or OC in microparticulate form has been used, all furniture should be placed outdoors for cleaning. Dry cleaning is suggested for decontaminating clothing and other fabrics; however, you should check with the dry cleaner first to ensure that it is willing and equipped to deal with the agents used.

All exposed foods must be discarded, including foods wrapped in plastic. (CN and CS can penetrate many plastics.) Canned foods may be saved, though the cans' exterior surfaces must be cleaned thoroughly before opening.

A HEPA-filtered industrial-type vacuum cleaner should be used to clean all furniture, carpets, draperies, surfaces, and floors. Standard household vacuum cleaners should not be used; not only will the machine be permanently ruined, it will only stir up the dust particles.

Surfaces that won't be damaged by moisture can be cleaned with a 5-percent solution of washing soda (sodium carbonate) in water. This solution helps decompose CN. Baking soda (sodium bicarbonate) may also be used, but it works more slowly. Nonionic cleaning detergents (e.g., Tide) can also be used when cleaning with water.

If the first attempt at decontamination is less than successful, you should repeat the process. Failing that, you may try the following procedure.

First, shut all doors and windows and heat the building as much as is practical (at least four hours at a minimum of 95°F (35°C).

Once the desired temperature has been reached, open a window at each end of the building. A large fan should then be placed inside at one of the windows, faced to blow outside. Leave the heat on. The heat should vaporize any CN or CS particles, and the fans will (hopefully) blow the vapors out the window.

Each of these cleaning techniques should be repeated until the building is free of contaminants.

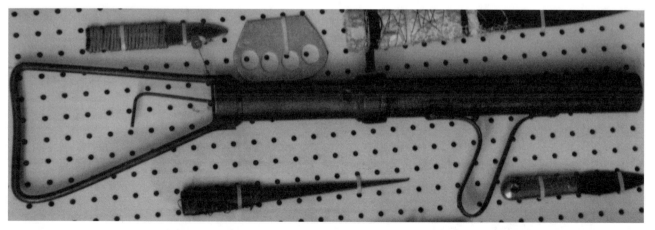

It is imperative that all canisters, casings, and munitions be recovered after an operation is completed. The large device shown in this photo (center) is a crudely constructed 37mm launcher made from pipes and a metal rod. Seized by corrections officers shortly after a prison riot in the 1970s, this weapon was fabricated by inmates to fire a 37mm round they had assembled. The round used a 37mm casing left behind by police after the riot and was loaded with homemade gunpowder, primer cap, and a fragmented scrap metal payload.

Protective Masks

HISTORY AND DEVELOPMENT

As noted in Chapter 2, most of the chemical agents we use today have existed in one form or another for many years. This is no less true with protective masks.

One of the first known chemical protective masks appeared in the 16th century. Leonardo da Vinci instructed sailors to cover their noses and mouths with a fine cloth dipped in water to resist the effects of a toxic powder weapon he had designed.[1] Simple masks such as these are still being used today by protestors and prison inmates to counter the effects of riot control agents.

The first patent for what would eventually evolve into the type of protective mask we know today was issued to Lewis P. Haslett of Louisville, Kentucky, in 1849. Described as an "inhaler" or "lung protector," it was designed for "protecting the lungs against the inhalation of injurious substances." The filter Haslett employed was made of wool or other porous substances moistened with water.

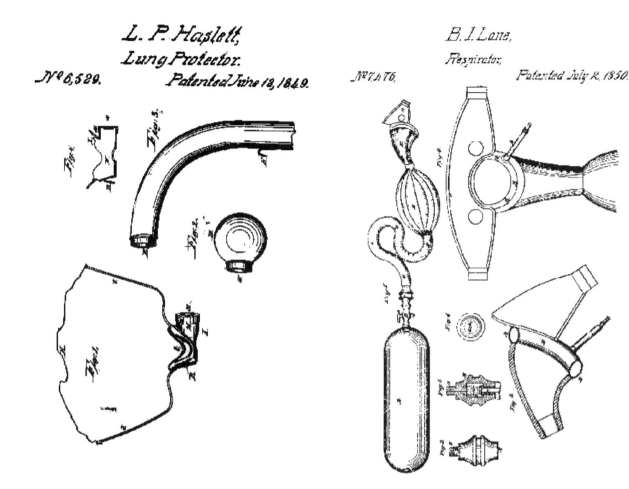

Illustration 18-1. Patent illustration of Lewis P. Haslett's inhaler, 1849.

Illustration 18-2. Patent illustration of Benjamin I. Lane's Pneumatic Life-Preserver, 1850.

One year later, Benjamin I. Lane of Cambridge, Massachusetts, patented another protective mask that he called the Pneumatic Life-Preserver. Its purpose was to allow people to enter "buildings and vessels filled with smoke or impure air and into sewers, mines, wells, and other places filled with noxious gases or impure air." Unlike Haslett's mask, Lane's was made out of vulcanized rubber and incorporated goggles into the design. Lane's mask was, in effect, the prototype of the modern chemical agent respirator.[2]

A British inventor named John Stenhouse produced another mask of note in 1850. Stenhouse's half-face, or oronasal mask (Illustration 18-3),

consisted of a velvet-lined facepiece covered by a filter element made with wood charcoal. The mask was held to the face by elastic headbands. Goggles were not attached to the mask.

Stenhouse, who first prepared chloropicrin (PS) in 1854, donated his mask to the public, and many of them were produced and distributed by chemical manufacturers to their workers. In addition to its resemblance to modern protective masks used in hospitals, laboratories, and numerous other contaminated-air environments, Stenhouse's mask is also strikingly similar to the present-day Riot Facelet Protective Mask (excluding the goggles) offered by AERKO International (see photo on next page).

Illustration 18-3. Archive, artist unknown.

Riot Facelet (RF) mask and goggles.

THE RIOT FACELET MASK

(Al Pereira photo.)

The Riot Facelet (RF) mask produced by AERKO International is made up of a charcoal cloth facelet secured over the nose and mouth by a snap-on elastic strap harness and a sturdy set of "gas tight" [AERKO terminology] goggles that mount over the nosepiece of the facelet and protect the eyes. The facelet itself is made up of an activated charcoal cloth, developed by the British Chemical Defence Establishment, Porton Down.

Advice in informational literature suggests that the mask piece be disposed of after use in contaminated areas. This would significantly increase the per-incident cost of the unit in contrast to conventional respirators. The goggles are made up of scratch resistant acetate. The RF Mask, described as a "civil derivative of the military NBC Facelet which is approved to NATO standards," is fairly effective. The one I tested, however, just did not impress me as being equal in performance or cost-efficiency to a standard chemical agent respirator.

Though billed as a low-cost alternative to such conventional respirators, the unit (costing approximately $70 complete with facelet, harness, and goggles) seems fairly limited in most tactical applications, for not only does the seal between face and facelet tend to be less than stable while worn, but the "adjustable harness for universal fit" also tends to unsnap at inopportune moments. While I believe that there is a need for a compact portable mask, I don't believe that the RF Mask is the ideal solution.

Various configurations of respirators were continually produced and refined over the latter half of the 19th century. Improvements to both mask design and filtering materials were made both in the United States and abroad.

Soft rubber facepieces began to appear in mask designs as did sockets that held replaceable eyepieces in place and provided a gas-tight seal. Elastic webbing was used to hold the mask securely to the face. The first respirator to incorporate an exhaust valve was patented by Hutson R. Hurd in 1889.

Nearly all these early respirators were designed to protect firemen and others who worked in environments where they were exposed to dangerous chemicals and vapors.

Not until the advent of World War I did the practice of using chemicals as weapons of modern warfare become fully realized. At the time, the grim specter of this new reality prompted the poet Thomas Hardy to observe wryly: "After two-thousand years of Mass; we've got as far as poison gas."

The Germans were the first to employ chemical weapons against Allied troops (primarily French and Algerian the first time) with deadly chlorine gas at the Belgian town of Ypres in 1915.

Despite the fact that protective masks were widely used in civilian industry at the time, the Allied soldiers had none and were caught totally unprepared for the vicious chemical weapon attacks. One of the first desperate responses of British commanders on the battlefield when their troops came under attack was to order their troops to "piss on your handkerchiefs and tie them over your faces." Shortly thereafter, untreated cotton mouth pads were hastily provided to the soldiers, with instructions to keep them moist to filter out the chlorine.

Later, women's yard-long black veils were issued to the troops along with the mouth pads. The pads were coated with sodium carbonate, sodium thiosulfate, glycerin, and water, and secured over the mouth and nose with the veil. The protection afforded by this field remedy was reportedly less than satisfactory.

During the summer of 1915, the British developed another simply designed mask, this one comprising a flannel bag soaked in sodium thiosulfate.

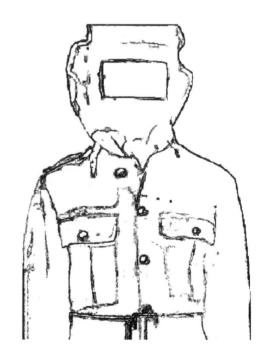

Illustration 18-4. Hypo ("H")-Helmet.

Dubbed the "H" or Hypo-Helmet (Illustration 18-4), the first version of the bag was designed with a mica viewing port that allowed the wearer to see. In the event of a gas attack, the bag was simply pulled over the head. It had no exhaust valve but did prove effective against chlorine. An improved version of this mask appeared shortly thereafter. The improved version was treated with sodium phenolate, glycerin, and later hexamethylenetramine, making it effective against phosgene gas as well as chlorine. An outlet valve was also incorporated into the new version, and later it would be improved again by the addition of goggles. The bag-type respirator design is still being used in commercial and military industries, though modern versions are much improved. One such respirator, dubbed the Quick Mask, was introduced in the 1990s.[3] Unlike the 1915 model, the modern Hypo-Helmet incorporates highly refined, fire-resistant materials.

QUICK MASK

(Al Pereira photos.)

Quick Mask is made in Israel and distributed by Fume Free, Inc. This lightweight (7-ounce, or 200-gram) emergency mask fits into a pocket-sized, sealed-foil carry pouch. It uses a seamless one-piece molded hood with a permanently bonded visor. All materials are fire resistant. The filter contains layers of charcoal cloth to absorb toxic gases and a particulate filter to remove harmful particles.

The sample I tested worked well. Donning it properly takes a little getting used to, and having the mask seal around the neck is definitely different, but, all in all, it works as advertised and once on is fairly comfortable. Communicating through the mask is also accomplished relatively easily because the thin layer of material directly behind the exhalation valve and in front of your mouth does not significantly impede your ability to be heard.

Described in company literature as a single-use mask for emergency escape, it is priced around $100. Though more expensive than the Riot Facelet Mask, the Quick Mask is faster and easier to put on and stays on while you're running and gunning.

Late in 1915, the Germans introduced a respirator with a design far more advanced than the Hypo-Helmet. This mask, known as the *Gummimaske* (rubber mask) or *Seidenstoffmaske* (silk material mask), actually served as the prototype for numerous mask designs that followed, including many that are available today. The mask incorporated a removable canister that attached directly to the facepiece by way of a threaded metal plate. The designers also incorporated the reverse face seal, which would prove exceptionally effective at maintaining a good seal between face and mask. Another clever feature involved the addition of extra material into the facepiece design on each side of the celluloid eyepieces. These triangular pieces of material, which can be seen in the **photo below**, were provided so the wearer could push his finger into the mask and wipe off the eyepiece without having to remove the mask if the eyepiece fogged up.

In early 1916, the first British respirator with a separate filter element appeared on the battlefields.

This development was in response to the Germans' introduction of new chemical warfare agents: it was believed that having the ability to change filtering elements would be more efficient and economical than having to change the entire mask. The separate canisters were attached to the facepiece by a hose, and required the wearer to use a nose clip and breathe through a rubber mouthpiece. Later British and U.S. versions would see improved facepiece materials and filtering elements.

The design of the Akron-Tissot (A.T.) mask of 1918 (Illustration 18-5) eliminated the uncomfortable mouthpiece and nose clip. Instead, air drawn through the filtering element and hose was directed over the eyepieces, thus helping to prevent fogging.

Respirators of this general type and configuration (though continually developed and refined) dominated U.S. mask designs up until 1944, when the M5 combat mask was introduced. The M5 Assault Gas Mask (Illustration 18-6) used a design that eliminated the hose between mask and filter. Instead, a lightweight canister that provided protection similar to that of the heavier canisters was attached directly to the facepiece, much like the German *Gummimaske* of 1915.

A third mask/filter variation was produced in

German soldiers pose for a photograph circa 1916. They are wearing the rubber *Gummimaske*. The mask was initially carried in a cylindrical bag that hung from the belt. The bag was later replaced by a metal carrier. (Photo from author's collection, photographer unknown.)

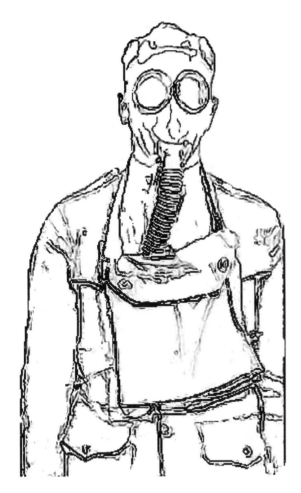

Illustration 18-5. The A.T. mask.

Illustration 18-7. M17 Field Protective Mask.

Illustration 18-6. The M5 Assault Gas Mask.

1959 when the U.S. Army standardized the M17 Field Protective Mask.[4] This mask eliminated the need for an external canister by placing the filter elements within the cheek pockets of the mask itself. A "voicemitter" was added to improve speech transmission, and the mask was produced in three sizes. The mask was refined in 1966 by the addition of a drinking tube and a resuscitation tube and designated the M17A1. In 1983 another version of the mask was produced with no resuscitation tube. Standardized as the M17A2, this version was also offered in a new extra small size.

Full-face mask designs based on these three basic configurations continue to be produced and refined. The primary differences between individual masks are found in materials and layout. The primary purposes of any full-face protective mask, however, have always been and still are to protect the wearer's eyes from contamination and to purify available oxygen through its filtration system.

MASK CHARACTERISTICS

With few exceptions, the full-face individual-agent protective mask available today is a self-contained unit that houses all necessary components in an integrally designed package.

Facepieces are generally made of rubber or other similar material that easily forms an airtight seal between the facepiece and the skin of the wearer.

The facepiece is normally kept securely in place by a series of straps worn around the head and attached to points around the circumference of the facepiece. A netlike harness may also be used this way.

Some kind of viewing portal is provided, most commonly in a dual-"eyelet" or a full-lens configuration.

Air is inhaled through a filtration system that is built into the facepiece, attached directly, or attached by way of a flexible hose.

Positive and Negative Air Systems

Two different systems are employed to allow clean air into the facepiece and are generally referred to as positive air systems and negative air systems.

A positive air system employs electric or mechanical blowers or pressure to lightly blow fresh air across the face of the wearer. This system is commonly found in specially equipped aircraft, tracked vehicles, and facilities; however, portable, individual, stand-alone units are also made.

A negative air system is found in the majority of standard masks. This system requires the wearer to literally "pull" clean filtered air into the mask through a filter as he inhales.[5]

NEGATIVE AIR SYSTEM RESPIRATOR AIR-FLOW PATTERNS

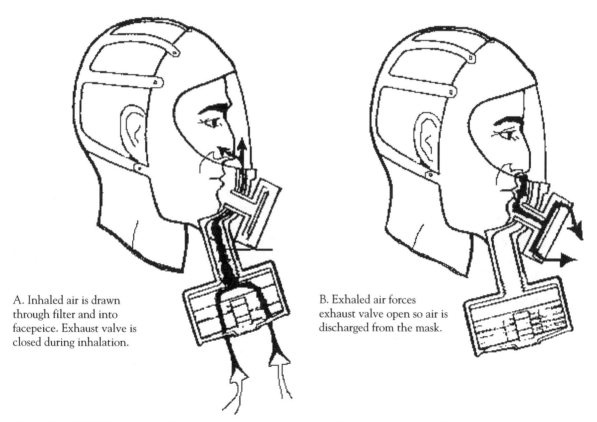

A. Inhaled air is drawn through filter and into facepeice. Exhaust valve is closed during inhalation.

B. Exhaled air forces exhaust valve open so air is discharged from the mask.

Illustration 18-8. This diagram illustrates the negative air system air-flow patterns during inhalation and exhalation.

Regardless of configuration, filter elements are usually replaceable or capable of being cleaned. Various types of filtering materials are used to remove particles and absorb irritant vapors from the air.

After the air passes through the filter, it is most often allowed inside the facepiece by a one-way inlet valve that seals when air is exhaled. Exhaled air then escapes through a one-way outlet valve that seals when inhalation begins again.

Vocal communications are also allowed with most masks. This is usually accomplished through an attached mechanical voicemitter or an electronic microphone.

Most masks are also compact enough to allow them to be carried in a carrying case that can be strapped or secured somewhere on the user's person until it is needed.

Some examples of the three basic configurations are presented on the following pages.

Remember that unless they have been equipped with some sort of self-contained breathing apparatus that provides a separate air source, protective masks of this type are designed only for use in agent-contaminated areas—they are not for use in atmospheres containing less than 19-percent oxygen or for entry into atmospheres that are immediately dangerous to life and health. The primary purposes of any chemical agent mask or respirator are to protect the wearer's eyes from contamination and to purify available oxygen through its filtration system. *They do not provide a separate oxygen source!*

TYPE 1 MASK:[6] CANISTER ATTACHED TO FACEPIECE BY HOSE

THEN . . .

Corrected English Model

The Corrected English Model mask (CEM), circa 1917. The CEM was produced by the Americans as an upgrade to the American Small Box Respirator and prior to the introduction of the A.T. Mask of 1918. It used celluloid eyepieces held in by metal rims. This mask is shown with the "H" filter canister attached. Also shown is the British-style haversack it was issued with and a card to record gas exposures. (Mask from author's collection.)

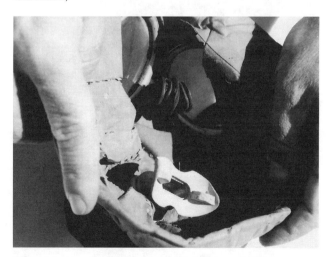

This view shows the snorkel-like mouthpiece and rubber nose clip. This design, common in many early respirators, was especially uncomfortable when worn for long periods. (Mask from author's collection.)

M2 Series

U.S. Navy Mark IV

Author's father, Jim Conti, training with an M2 Series box-type respirator during World War II service. (Photo from author's collection, photographer unknown.)

World War II era U.S. Navy Mark IV gas mask, front view. This mask was unique in that it used two hoses that connected to a single filter element. (Mask from Marty Driggs' collection.)

M2 series mask. These masks became obsolete in 1949. (Mask from Marty Driggs' collection.)

U.S. Navy Mark IV mask, viewed from above. Canister could be worn in front or behind neck. (Mask from Marty Driggs' collection.)

. . . AND NOW

M45

Illustration 18-9. The M45 Chemical-Biological Mask was placed into service in 1996. Designed for use by both infantrymen and military aviators, the mask is compatible for use with the complex sighting and night-vision devices used in most rotary-wing aircraft. The mask is currently produced in four sizes and provides protection without the aid of forced ventilation air as was required with the earlier M43 series masks.

U.S. M1A2

U.S. M1A2 noncombatant mask. This mask was produced with a rubber-coated stockinet facepiece and plastic eyepieces. A lightweight, nonreplaceable, cylindrical canister was attached directly to the facepiece. More than 6 million of the M1A2 masks were procured during the war. The M1 Series masks were obsoleted in 1954. (Mask from the author's collection.)

TYPE 2 MASK: CANISTER ATTACHED DIRECTLY TO FACEPIECE

Three examples of type 2 respirator designs, clockwise from bottom left: the Advantage 1000, Model 6006, and Phalanx Alpha. (Photo courtesy of Federal Laboratories.)

Scott Respirator

Many type 2 respirator designs are available. Some, like the Scott Chemical Agent Respirator and others of similar design originally intended for industrial use, have been adopted by various police agencies throughout the world. All quality-made respirators will perform their intended function, but some are better suited for the particular rigors of police work. While the full-facepiece lens does provide an excellent field of view, it also can fog up to varying degrees. This is especially likely when the mask is worn under stressful or physically active conditions—the same conditions that are almost ensured when equipment is needed for tactical applications. The absence of some form of nose cup is probably the cause of the fogging. I was issued this kind of mask in 1986, and, like many masks over the years, it came complete with an antifogging agent in a small plastic dispenser. The agent, applied as directed, worked for a short time but, in my experience at least, never prevented the mask from fogging in real-world tactical conditions.

To give it its due, the mask itself, equipped with a nylon-mesh head harness, is fairly comfortable to wear and has never failed to function properly. The fogging problem and the long physical profile caused by the forward placement of the canister have, however, proven detrimental during both training and actual operations.

The filter element on this and the other type 2 masks shown in this section are easily replaced simply by unscrewing the old and screwing on the new. This should not be done in contaminated environments for obvious reasons.

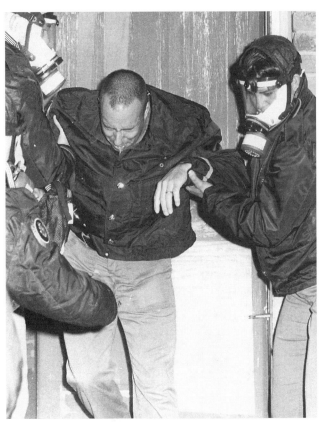

Instructors equipped with modified Scott Chemical Agent Respirators assist trainee from "gas house" after controlled exposure to agents. (Mark C. Ide photo.)

Israeli Civilian Respirator Model 4

Many of these Israeli civilian respirators, produced by Shalon-Chemical Industries, Ltd., have been exported to the United States and other countries over the past few years. Extremely cheap, the masks can *usually* be purchased complete with an extra filter element and a carrying case (normally surplus U.S. M17-type carrying cases) for around $20.

You do, however, get what you pay for. Most of these masks are at least 10 to 20 years old, having been sitting in warehouses gathering dust, and are approaching their expiration dates. They have small eyelets and are not very comfortable to wear, especially for extended periods. In my experience, the seal between the face and mask may also be easily broken when the wearer is running or involved in other strenuous activities. Make no mistake about it, though; the mask is effective, at least against CAWs.

Scott Respirator. Seals should be left on the filter until the mask is used. (Mask from the author's collection.)

Israeli Civilian Respirator Mask Model #4. (Mask from the author's collection.)

An untold number of these respirators have been sold in this country, many at ridiculously inflated prices, since the despicable terrorist attacks against the United States in September 2001. Bottom line: Due to these shortcomings, this particular mask is not recommended for serious law enforcement or military use.

Advantage 1000

The Advantage 1000, produced by Mine Safety Appliances (MSA) International of Pittsburgh, Pennsylvania, is one of the best tactical-applications respirators I've ever encountered. Its history can be traced back to the XM29, XM30, XM33, and XM34 series of masks developed and evaluated by the U.S. Army during the 1970s and 1980s. Although never standardized by the army, the U.S. Air Force did adopt it and completed its development as the MCU-2/P. Both the U.S. Air Force and Navy employed the mask during Operation Desert Shield and Desert Storm. The primary differences between

The Advantage 1000, shown here without and with tinted or smoke-colored polycarbonate "outsert" cover lens in place. This accessory adds tremendously to the psychological impact of the mask while allowing a clear, sharp view of the world for the wearer. A clear cover lens is also available. These lens shields may be secured to and removed from the mask in seconds. Once in place, they appear to be an integral part of the mask and not some type of aftermarket add-on. They provide a greater degree of added impact protection for the wearer as well as scratch protection for the integral urethane lens. (Mask from the author's collection.)

the military version and the Advantage 1000 are that the military model has a drinking tube similar to that found on the M17A1 and accepts NATO-standard threaded canisters. The Advantage 1000 has neither of these components.

I first had an opportunity to examine the Advantage 1000 while participating in a chemical agent instructor program in 1996. I was so impressed by both the design and function of the mask that I wrote an article for *S.W.A.T.* magazine reviewing it.[7]

The Advantage 1000 not only has a distinctive, psychologically striking appearance when worn with the tinted lens shield, but it is also the most comfortable, least intrusive mask I have ever worn. This opinion is based on my experiences in using various types and configurations of masks in both the military and police over the past 20 years.

It is simply designed and cleanly contoured to minimize interference with the wearing of the standard Kevlar protective helmet. A single filter-element canister provides sufficient protection against agent-contaminated dust, fumes, and mists, and may be worn on either the left or right side. Masks that incorporate this feature allow the user, whether right- or left-handed, a significant advantage when firing shoulder-mounted weapons. Two canisters may also be used simultaneously to increase exposure time.

The flexible, integral urethane lens and black Hycar rubber[8] facepiece also allow for an extremely comfortable fit as well as seal.

The mask is very lightweight and resists fogging quite well. This is most likely attributable to the mask's removable nose cup. A spectacle kit is available for those who wear corrective lenses.

Phalanx Alpha Gas Mask

The Phalanx Alpha Gas Mask, also produced by MSA, has a one-piece clear urethane lens similar to that of the Advantage 1000. The mask is very effective and comfortable, and designed to be worn with the standard Kevlar protective helmet.

A single external-filter element canister provides sufficient protection against agent-contaminated dust, fumes, and mists and may be worn on either the left or right side. This allows for faster filter changes

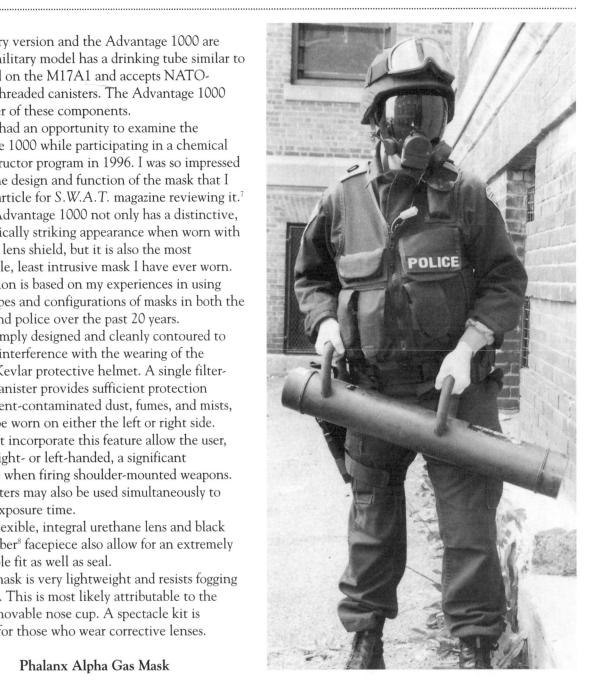

An Everett, Massachusetts, Police Department special operations officer (above) wearing the Advantage 1000 with electronic voicemitter. The battery-operated microphone and speaker system (ESP, Communications System) is housed in a waterproof, injection-molded, high-strength polymer case. It can be attached in place of the mask's integral voicemitter, significantly increasing the range of vocal communications. This feature proves extremely useful for command and control purposes, especially in confined environments where there is significant background noise. Such an environment may be created when smoke detectors are triggered by distraction devices during dynamic raids.

The ESP Communications System attached to the Advantage 1000. This version is equipped with a red light-emitting diode (LED indicating power on (visible above battery compartment screw on right). This light should be disabled or covered when using masks operationally.

The Advantage 1000 is extremely flexible and comfortable to wear.

Phalanx Alpha Gas Mask. (Photo courtesy of Federal Laboratories.)

and also makes it easier to employ shoulder-mounted weapons while masked. Two canisters may also be used simultaneously to increase exposure time.

The mask is also very lightweight, and resists fogging quite well. A spectacle kit is available for those who wear corrective lenses.

M-95 Tactical Mask

Another entry in the tactical respirator field is the excellent M-95. This Finnish-made mask is one of the highest quality gas masks available in the world today. It can be used with great comfort while wearing the ballistic helmet. The M-95 uses hardcoat polycarbonate lenses and outserts held in place with sturdy metal rings. It comes equipped with an integral drinking valve. A spectacle kit is also available.

The M-95 is currently being adopted by many police departments and military organizations throughout the United States and Europe.

TYPE 3 MASK:
FILTER ELEMENT
WITHIN THE FACEPIECE

M17A2 Mask

The M17A2 mask. The primary difference between the M17A1 and the M17A2 is that the M17A1 has a resuscitation tube and the M17A2 does not. Note the drinking tube below voicemitter. (Mask from the author's collection)

Internal view of the M-95 shows nose cup and integral drinking valve. Special water canister and connecting tubes allow the wearer to replenish fluids while masked. This type of component is most commonly found on masks intended for military use.

The masks in the M17 series were extremely well made and reliable. Many law enforcement officers who served in the military have used one or more of the M17 variants. As a result of this familiarity and the fact that these military masks have been made available to civilian law enforcement agencies through U.S. government programs, the M17 series mask has been extensively adopted for civilian law enforcement use in the United States.

NOTE: The U.S. Army has replaced the M17A2, however, with the M40 series masks, and the available M17A2 inventories and components are aging. This aging may result in deteriorated facepieces, and the preservative may have lost its effectiveness while in prolonged storage.

A couple of other facts regarding the M17 series masks are also of critical importance to any U.S. civilian law enforcement agency currently using (or thinking of adopting) the M17:

- MSA, the principal manufacturer of the M17 mask under contract with the Department of Defense until production was terminated in the 1970s, does not recommend this mask for use by civilians.
- The M17 is not approved by the Mine Safety and Health Administration and the National Institute for Occupational Safety and Health under the provisions of 42 CFR 84.

Accessories

Many specialized accessories were produced for use with the M17 series masks. The military versions were generally equipped with a chemical/biological (CB) agent protective mask hood[9] that attached to the facepiece itself and was pulled over the head when the mask was donned. For police applications, these hoods may prove more a hindrance than a help.

A winterization kit, designed to be used in cold weather, was also available. The unit consisted of a

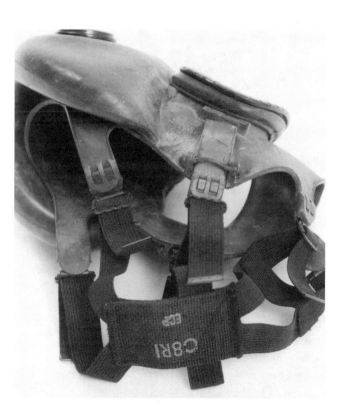

Rear view of the M17A2 showing head harness. U.S.-made M17 series masks should be checked for deterioration before using because existing supplies are aging. (Mask from the author's collection.)

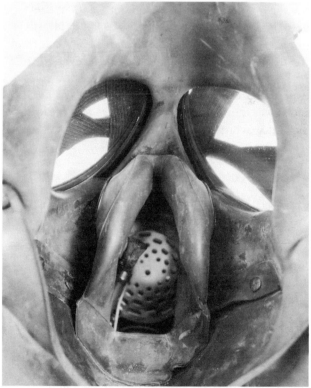

Internal view of the M17A2 shows nosecup and integral drinking valve. Voicemitter plate can also be seen. When it is being donned, the chin should be placed into chin cup visible at bottom of photo. (Mask from the author's collection.)

prefilter and a set of special inlet valve disks and nosecup valve disks that prevented frost accumulations on the inlet valves when operating in extreme cold.

Optical inserts were available for those who need prescription lenses. Two types were available for use with the M17, however, only one of them could be used with the M17A1.

A resuscitation tube was also made for the M17A1 variant. This tube, which attached only to this variant, allowed the wearer to perform rescue breathing while masked in a chemical environment.

Changing Filter Elements

Changing the M17 series mask filter elements is neither quick nor easy. This is one of the biggest drawbacks with respirators configured with internal filter elements. Changing filter elements requires the mask to be turned inside-out after removing the inlet valves and releasing the filter element connector flange. The nose cup must then be unbuttoned from the flap. The top pouch flap on both sides of the mask must then be unbuttoned. Only after all this has been done can the filter elements be pulled from the mask. The procedure must then be reversed to replace the old elements with the new. In my experience, the best part of changing the M17's filters was found in the variety of new and creative profanity overheard when working on the masks with a large group of people.

THE FUTURE . . .

XM50 Joint Service General Purpose Mask

The XM50 Joint Service General Purpose Mask is described by proponents as a revolutionary advancement in protective mask technology. The XM50 is currently being developed for all services to replace the army's M40 and M42 series masks, as well as the air force's and navy's MCU-2/P mask.

The mask reportedly provides significantly enhanced protection against toxic industrial materials as well as reduced breathing resistance. Combined with an improved field of view and reduced weight, the XM50 should provide a distinct advantage for the members of the U.S. military well into the 21st century.

Illustration 18-10. The XM50 mask.

GENERAL DONNING PROCEDURE

The unique design and harness configurations found on the various makes and types of protective masks require you to receive specific training regarding each mask's individual care and use. This information should be made available to you with any new mask you acquire. If not, then it is recommended that you seek out experienced and competent instructors for advice and training.

NOTE: The following section illustrating one method of donning a protective respirator is provided for general information purposes and may not be appropriate for other types or configurations of masks.

2. Grasp the facepiece with both hands, thumbs inside. Open the mask fully. Place chin into the chin cup.

1. STOP BREATHING! This is a good habit to get into when donning a mask because it prepares you for the possibility of being exposed to agents unexpectedly. Then remove any headgear you may be wearing. If carrying a long gun, you can place the helmet on it as it is secured between legs. If you are not, place helmet between legs (if possible) to make sure the helmet is not lost or forgotten. (For military personnel, this practice also minimizes contamination to the helmet should you find yourself operating in a toxic-agent-saturated area.) Then remove mask from carrier. The harness should be adjusted so that it is loose enough to fit comfortably over your head.

3. Pull the head harness over your head. Straighten the straps and adjust the mask quickly, smoothing the edges of the facepiece back and upward with the hands.

4. Sharply tighten the straps (if necessary) in a series of twos, beginning with the bottom . . .

5. . . . and working up to the temple and head straps.

6. Cover the outlet valve cover with hand and blow out hard. This will clear any agent from the inside of the mask.

7. Check seal by placing hand (or hands, depending on the number of filtering canisters used) over the intake port(s) of the canister(s) (or inlet valves for type 3 mask) and breathing in. If the seal is good, the mask should collapse around the face and no air should enter around the facepiece. If air does enter from around the mask, readjust and repeat the clearing/sealing procedure. After seal is established secure hood over head if mask is so equipped and replace headgear.

8. Chemical agent hand and arm signals can be used to alert other officers of presence of CAWs. Upper arms are held as shown, parallel to deck, forearms at 90-degree angle to upper arms. Fists are then pumped vigorously toward shoulders and back to 90-degree angle. Verbal warning of "Gas!" may also be used, AFTER you have masked.

USE

It must be remembered that all the chemical protective masks shown in this section are designed only for use in agent-contaminated areas. These masks serve to protect the eyes from contamination and to provide purified *available* oxygen through their filtration systems. *They do not provide a separate oxygen source!*

This fact must not be misunderstood. If you enter an environment where the oxygen has been overwhelmingly displaced by what has been contaminated by corrosive toxins or poisons, the chemical agent protective mask will be of little or no help to you.

Occupational Safety and Health Administration (OSHA) Rulings

OSHA has several regulations pertaining to the industrial use and wearing of chemical respirators. They are noted here for information purposes.

Rule 29 CFR 1920.134 (e) (5) (i)

- Respirators shall not be worn when conditions prevent a good face seal. Such conditions may

GENERAL DOFFING PROCEDURE

1–2. After ascertaining that the area is clear of contaminating agents, remove the mask by grasping it under the chin area with both hands and pulling it up and over the head. If equipped with a hood, the hood should be removed first by pulling it forward and over the head and mask. (Small unit police and military tactical intervention team members should assess whether the hood is necessary for their particular applications. If not, you may want to omit it.) Mask should then be returned to carrier and secured. Once you are unmasked, the "all clear" signal may be indicated by verbal command. If someone other than you sounds the signal, you should verify that he is both in your immediate area and unmasked before removing your own respirator.

be a growth of beard, sideburns, a skull cap that projects under the facepiece, or temple pieces on glasses.[10]

Rule 29 CFR 1910.134 (e) (5) (ii)

- Non-gas-permeable hard contact lenses may *not* be worn under mask.[11]
- Gas-permeable and soft contact lenses *may* be worn under mask.[12]

Special Use Considerations

Night-Vision Goggles–Chemical Agent Mask Combination

Night-vision goggles with mask.

Occasionally it may be either necessary or desirable to enter a contaminated area when it is dark. The obvious solution to this operational need would be to use night-vision goggles (NVGs) in conjunction with a chemical agent respirator. Should there be known or potential adversaries in this environment, however, one minor (potentially major) problem exists: when using standard NVGs with a respirator, the eye cups of the goggles will not seal around the user's eyes. Rather, the cups will be held away from the eyes and face of the wearer.

Although the NVGs enable the wearer to see in

Single-lens NVG with mask.

the dark, the green light emanating from the goggles may also illuminate his face to his adversary. The effect is more pronounced when wearing fully lensed respirators (e.g., the Advantage 1000) because more of the operator's face is exposed in the glow. The effect is most noticeable when the wearer's face is seen in profile.

This has to be considered before you use the respirator and NVGs together; otherwise, you or your operators may be faced with an enlightening predicament.

Oxygen Capability

Although the masks in this section are shown equipped only with filtering canisters suitable for agent-contaminated environments, some respirators do allow for entry into environments where the oxygen has been depleted or displaced.

176

MAINTENANCE

Inspecting Mask and Carrier

1. Remove mask from carrier. Inspect all components. Immediately report any missing components to your supervisor or chemical agent munitions officer.
2. Check the carrier for wear, dirt, missing straps, or broken hardware.
3. Check the mask for holes, wear, splits, rubber fatigue, dirt, or mildew.
4. Check all filter elements. Ensure that they are serviceable. Replace any filter elements that are wet, have been directly exposed to water, or have passed their expiration dates.
5. Check eye lenses for scratches, cracks, or discoloration.
6. Check head harness for wear; broken or frayed straps, dirt and mildew, or broken strap securing hardware.
7. If mask is equipped with a hood, check it for wear, tears, holes, dirt, or mildew.

Cleaning Mask

1. Remove voicemitter outlet covers, inlet valve caps, and protective lenses or outserts.
2. Remove external filter elements. (On type 3 masks, filter elements may generally be left in for routine inspection and cleaning.)
3. Clean mask inside and out with a soft cloth that has been dipped in warm, soapy water. Cloth should be wrung out thoroughly before use. A soft-bristle brush may also be used. *Be sure not to wet the filter elements!*
4. Rinse with a soft cloth that has been dipped in warm, clear water. This cloth should also be wrung out thoroughly before use.
5. Dry with soft, lint-free cloth, or allow mask to air-dry.
6. Reassemble mask.

NOTE: Many chemical agent protective masks have a light-white or blue-tinted powder on the rubber components. This is a rubber preservative and should be left on. No alcohol or alcohol-based products should be used to clean rubber components because the alcohol will damage the rubber and may cause toxicity.

Replacing Filter Elements

1. Filter elements should be replaced:
   — after exposure to chemical agents
   — after being immersed in water
   — after being damaged
   — after being used by someone with a respiratory illness

Clean Carrier

1. Empty carrier completely.
2. Brush off the carrier inside and out, removing any dirt, sand, or grit.
3. If the carrier is soiled, clean it with a brush and cold water.
4. Reassemble carrier components.

The SF10 Respirator produced by RBR Armour, Ltd., in the United Kingdom is an example of a type 2 mask that has been designed with this capability, in addition to providing protection against CS, CR, and other irritant gases, aerosols, and smokes.

MAINTENANCE

The chemical agent protective mask must be maintained in serviceable condition at all times. Regularly scheduled, routine servicing of the mask ensures that the equipment will not fail when needed most. This servicing is critical and must be performed, especially if your unit or team does not routinely train with or employ chemical agents. The protective mask, like most gear that is normally out of sight/out of mind, seems to be especially susceptible to unnoticed damage. It is not unheard of for a mask to be pulled from its carrying case only to show that somewhere along the way a lens has been cracked or a filter contaminated with water. It is also very important that training be conducted regularly in actual chemical-agent-contaminated environments. If this is rarely or never done, then operators may fail to learn how to properly and quickly don and effectively seal their masks to their face.

This need for preventive maintenance and continual agent-exposure training was dramatically illustrated one afternoon when several of my teammates and I performed an entry into an agent-contaminated apartment during a call-out. The offender, who had been holed up for hours with several firearms and a few hundred rounds of ammo (a dozen of which he fired out the window before surrendering), had been persuaded to leave his hooch by a thorough application of CS and OC agents. After he was safely cuffed and stuffed, the apartment had to be cleared to ensure that there were no other bad guys, hostages, victims, or dangerous items inside.

Forming up quickly beside the door, officers donned and adjusted masks as the silent countdown began. The lead man then entered, with the rest following close behind. As we swept through the house in dynamic fashion, moving fluidly from room to room, I suddenly observed first one and then another of my teammates as they apparently "flamed-out" in front of me, their masks allowing a healthy dose of agent into their lungs and eyes.

They were quickly advised to extract themselves, and those who remained finished the rest of the house clearing. Later, we discovered that one of the masks had a small perforation that had gone unnoticed in a recent inspection and that the other had failed to seal properly against the officer's face.

Fortunately, no other threats were in the house, besides the ever-present and unpredictable Mr. Murphy.

STORAGE

Once thoroughly cleaned, dried, checked, and reassembled, the mask should be placed in its carrier and secured in a cool, dry area.

Care should be taken to prevent damage to the mask and carrier if it is to be kept in the trunk of a cruiser or an equipment locker.

If storing a mask for long periods, it should be placed in a plastic bag, secured, and then returned to the carrier.

There is no way to determine the true life of a filter, at least in regard to effectiveness against CAWs. Most masks and filter elements will last for at least 10 to 15 years if stored properly. (See below for more information.)

Protect from Humidity and Moisture

Excessive dampness and heat should be avoided in order to minimize deterioration of the rubber parts of the mask. Storing the mask and filters in a dry area with a moderate controlled temperature will also prolong the service life of filter elements.

It's also a good idea to leave the head straps loose so you can don the mask quickly should an emergency arise.

MORE ABOUT FILTERS

Filters are designed to protect the mask wearer from various types of agents. You must make sure that you have the proper mask-filter combination for the type of environment you'll be operating in. Not all filters provide protection against nuclear-biological-chemical (NBC) warfare agents. Many types of expired or improper filters are currently being offered to a nervous populace. Most of these filters provide little or no protection against nuclear or biological hazards, despite unscrupulous claims to the contrary.

As a general rule, you should do the following:

- Use only newly produced, high-quality masks and filters that have not reached their identified expiration dates.
- Be concerned that the mask and filter combination you use to protect yourself is rated for the types of agents you may be exposed to.

NOTES

1. This weapon was described by him as a shell loaded with arsenic and sulfur dust that was launched against a ship.
2. Lane would further improve his mask design during the U.S. Civil War, patenting a version that used a full facepiece in 1865.
3. Other military versions of bag-type respirators have been produced through the years, including the M7 headwound mask (1944) and a children's "bunny mask" that was produced and distributed throughout the Hawaiian Islands following the attack on Pearl Harbor.
4. Though often referred to as *gas masks*, the term is actually a misnomer. Protective agent masks simply filter out chemical agent particles that float through the air. A true gas will not be filtered out. When dealing with an air-displacing gas, some type of self-contained breathing apparatus will be required.
5. It must be noted that breathing efficiency while wearing the standard negative air system mask is reduced because air must be drawn through the filters. This may cause fatigue more quickly than usual if you need to perform strenuous activity while wearing the mask.
6. For clarity's sake, the three mask design types referred to in this section are classified as types 1 through 3.
7. "ADVANTAGE 1000: A Product That Lives Up to Its Name," *S.W.A.T.* (April 1997).
8. According to the company, Hycar rubber is a superior formulation similar in characteristics to silicone, which provides for maximum resistance to chemical permeation.
9. The ABC-M6A2 field CB mask hood.
10. Respirators should not be used in atmospheres containing less than 19-percent oxygen.
11. According to OSHA, "The issue with non-gas-permeable hard contact lenses will be resolved in the revision effort for 1910.134 which is currently under way."
12. "The use of soft lenses shall be documented in case file and recorded as *de minimus*; citations will not be issued for *de minimus* violations. Any evidence relating either benefits or negative effects associated with the use of contact lenses with respirators should be relayed to OSHA."

Maintaining the
Chemical Agent Arsenal

DETERMINATION OF VARIOUS ELEMENTS

The decision to acquire a chemical agent munitions inventory cannot be made lightly. There are many considerations that must be taken into account. Before anything else, the justification for these types of munitions must be established. This justification is based on three factors: need, commitment, and preparation.

Need

- Has your department encountered situations that required a chemical agent response?
- Does the potential exist in your community/jurisdiction for situations that will require (or justify) a chemical agent response?
- Will the community support your use of chemical agent munitions?

Commitment

After determining need, you must then establish a commitment to the procurement of, training with, and use of these munitions. This commitment must start at the top and include the entire chain of command as well as elected officials in the community. Without this commitment, you would be much better off spending the money on something else.

Preparation

Once both need and commitment have been recognized, you must prepare your department and officers.

- A policy must be developed and approved.
- A departmental chemical agent munitions officer must be selected and trained by a reputable organization or entity.
- A secure storage facility must be organized.
- Arrangements for the procurement and disposal of the agents must be made.

CHEMICAL AGENT MUNITIONS OFFICER

A chemical agent munitions officer (CAMO) should be selected from department personnel. Some departments assign this duty to the armorer. Other departments let their tactical operations people absorb this responsibility. Regardless of who is authorized to carry these agents and trained to employ them, one specific person should be given the responsibility for maintaining the arsenal, updating records, and ensuring that all the munitions are properly used and accounted for.

To assist the CAMO in fulfilling his mission, certain standards and procedures should be established.

- An inventory control logbook should be instituted and should reflect the existing chemical agent munitions inventory and be regularly updated as munitions are procured, used, rotated, and destroyed. The logbook also helps to ensure that all munitions are accounted for so that none are left behind at the scene.
- A secure storage facility should be either built, procured, or designated.

CHEMICAL AGENT MUNITIONS STORAGE ROOM

You do not need to expend a great deal of department funds and purchase a high-tech, steel-reinforced, state-of-the-art chemical agent munitions storage container to begin operations. A simple concrete, temperature-controlled, windowless room will provide a suitable storage facility to serve all your chemical agent needs quite nicely.

The storage room should be located in an appropriate location, preferably dry and secure.

Munitions should be stored off the floor in their original packing containers until needed. They should be segregated by type and stored in an orderly fashion.

Ideal storage conditions for chemical agent munitions require a consistent temperature from 60 to 75°F, although a temperature as high as 80°F might be satisfactory.

Humidity should also be controlled. A level of 30- to 35-percent humidity is preferred, though even as high as 65 percent may be acceptable.

Stored munitions should be inspected for deterioration and function-tested twice a year to confirm their reliability. Access to the storage room and the munitions themselves should be restricted to trained, authorized personnel.

AGENT/MUNITION SELECTION

Chemical agents and munitions should be selected with care and a great deal of forethought. Most departments or agencies are probably best served by selecting, purchasing, training with, and employing only a few different types. These munitions should be selected based on the department's specific needs.

At the least, it is recommended that the basic police load includes 12-gauge barricade-penetrating (ferret) rounds and indoor CAW grenades for barricaded suspect situations, and some outdoor CAW and smoke grenades for civil-disturbance control.

As far as selection, most major modern chemical agent companies produce similar types of munitions with only minor variations. This is not surprising when you consider the historical development of these devices and the fact that there are only so many possible variations in design and production. That almost all these devices are also simple to use and highly dependable reinforces the "if it ain't broke, don't fix it" approach to design and manufacture.

There are some notable major chemical agent munitions producers outside the United States, such as Schermuly in Great Britain and ISPRA in Israel, but two of the most well-known commercial producers are in the United States, Federal Laboratories and Defense Technology Corporation. Both companies produce some of the highest quality chemical agent products in the world.

Mace Security International used to be a major player and actually owned Federal Laboratories at one time. Then in 1996 a new conglomerate, Armor Holdings, Inc.,[1] was formed and purchased Defense Technology Corporation. Two years later Armor Holdings acquired Federal Laboratories and the Mace trademark. As of this writing, Armor Holdings holds the major chemical agent munitions cards in the United States and is undoubtedly the best source for acquiring most of the chemical agent products discussed in this book.[2]

BASIC LOAD PURCHASE

As noted above, the basic police load should include 12-gauge ferret rounds, indoor grenades, and outdoor grenades. Such items as gas launchers and launchable munitions and foggers are always nice to have on hand, but unless a real need has been determined for them you're better off saving the money and storage room for needed munitions. (You should, however, do some tests and evaluations and decide on an appropriate chemical agent respirator that meets your department's needs and budget.)

Usually, 12-gauge ferret rounds come packaged five to the box. Enough rounds should be bought to allow authorized officers to train enough to become proficient in their use. The amount needed depends on projected operational needs and the number of officers who will be trained and authorized to employ them.

Aerosol, "flameless," or other munitions for indoor use should also be purchased in quantities sufficient for training and operations.

Finally, standard pyrotechnic grenades, triple-chasers, or other high-volume-producing grenades for outside use should be purchased in quantities that will allow for training and use.

RESUPPLY AND INVENTORY ROTATION

For liability reasons, munitions should be replaced prior to their expiration dates. Shelf life for typical modern chemical agent munitions is generally four to six years.

Under ideal storage and maintenance conditions, munitions may retain their effectiveness for 15 years or even longer. The liability inherent in the use of old or outdated munitions, however, must be considered. For this and other reasons, it is considered prudent to rotate the chemical agent munitions stocks.

For example, if you purchased an initial basic load, consisting of the types of hand-delivered munitions described in the "Basic Load Purchase" section above, on 1 January 2002, then you might have an inventory control logbook entry similar to this:

Date In	Quantity	Type	Description	Shelf Life (Years)[3]
01-01-02	100	Def-Tec #2	CS grenade, outdoor	4
01-01-02	30	Def-Tec #T-16	OC grenade, indoor	5
01-01-02	30	Fed. Labs MPG	CN expulsion grenade	6

If you then rotate this stock yearly and replace only what you use, annual replacement costs will be lower, supplies will be kept current, and your personnel will have munitions available to train with. For example:

Date Out	Quantity	Type	Description of Use	Disposition
01-01-03	25	Def-Tec #2	Training, crowd control	25 used/25 recov.
01-01-03	6	Def-Tec #T-16	Training, barricade susp	6 used/6 recov.
01-01-03	5	Fed.-Labs MPG	Training, SWAT Team	5 used/5 recov.

DISPOSAL

Disposal of expired, damaged, or otherwise unserviceable chemical agent weapon munitions can be a problem in today's environment, which is highly regulated and policed by the Environmental Protection Agency (EPA). The Resource Conservation and Recovery Act (RCRA) of 1976—which set federal guidelines for the management, transportation, treatment, and disposal of hazardous wastes—established severe penalties for violations of its standards.

NOTE: Outdated or unserviceable chemical agent weapon munitions *are* considered as hazardous waste according to RCRA guidelines.[4]

Many states also have strict environmental laws governing the disposal of these materials. Methods that until recently were used to dispose of these devices may no longer be used without ascertaining their legality. For example, in earlier years, the recommended method for destroying old munitions was to incinerate or burn them. Incineration, involving temperatures in excess of 750°F, was considered the most effective way to ensure complete oxidation of contaminating materials. Burning the munitions in pits (or even barrels), while not as efficient as incineration, was another option commonly used. Pits would be dug, and roaring fires would be built in them. The unwanted munitions would be tossed in and burned thoroughly; the pits were then filled and forgotten about. Fifty-five-gallon steel drums were also used to burn munitions in this manner.

Today, the *only way* to absolutely ensure that your disposal of old chemical agent munitions does not become a huge liability problem for you or your department is to *have the problem handled by a federally approved hazardous waste management facility*.

This method is generally expensive, but not as costly as having to pay fines that can total many thousands of dollars. However, the process of locating a reputable, certified, and honest disposal company may be more difficult than you imagine.

This point was driven home to me while I was attending a chemical agent munitions school, when one of my classmates, who happened to be an EPA special agent, advised the class that one of the larger hazardous waste management plants in the country was being investigated for EPA violations. No one in the class was more surprised than the instructor—who had just recommended the firm for use.

And as the EPA special agent further advised us, even if you hand over your hazardous material (HAZMAT) in good faith to a waste management company, you are still responsible for it, should the company violate any federal or state regulations during its handling or disposal.

So before you hand over any HAZMAT to any firm, investigate both it and its background thoroughly. The strife you save may be your own.

NOTES

1. Armor Holdings, Inc., is headquartered in Jacksonville, Florida.
2. Armor Holdings also offers comprehensive instructor-level certification programs throughout the year.
3. Shelf life used in the examples may not be accurate.
4. These types of devices have been assigned the EPA hazardous waste code of D001 because of their incendiary properties.

Basic Training Guidance

INTRODUCTION

This section is intended to assist officers in the development of basic individual practical skills. For ease of use, I've structured this section in a format similar to that used for such U.S. military training manuals as field manuals (FMs), soldier training publications, and Army Training and Evaluation Program manuals.

This section can also be used as an immediate training reference source, assisting certified and experienced trainers to provide limited training modules for members of their unit, department, or organization.

NOTE: It is critical that every department using chemical agent weapons develop its own viable, documented training program. Initially, departmental trainers can receive formal instruction from other agencies or from the manufacturers that produce the products that will be used. A department-specific training program in accordance with the agency's policies and procedures should then be formatted.

PERFORMANCE OBJECTIVES AND TASK SUMMARIES

An index of performance objectives is provided below. Suggested primary tasks are numbered and identified in this index, and subtasks that fall under the primary task heading are also identified.

Each performance objective identified in this section is broken down into three components: task (the job at hand), conditions (what the officers have to work with and any limitations imposed on them), and criterion or criteria (the standard of performance).

Training guidelines that provide a good—but not the only—method of performing the task are also included in this section.

TASK 1:   AEROSOL SUBJECT RESTRAINT (ASR) SPRAYS
   A. Presentation
   B. Static application
   C. Dynamic application
   D. Recovery to carrier
   E. Transition drill

TASK 2:   SMOKE AND CHEMICAL AGENT WEAPON DEVICES
   A. Perform safety check on devices.
   B. Identify and deliver devices.

TASK 3: **CHEMICAL AGENT PROTECTIVE MASK**
A. Perform operator's maintenance.
B. Don, clear, and test seal.
C. Wear mask in chemical environment.

TASK 4: **INDIVIDUAL OPERATION SKILLS**
A. Deliver smoke or CAW to designated OA in tactical environment.

TASK 5: **UNIT OPERATION SKILLS**
A. Operate in CAW-contaminated environment.

TASK # 1 – AEROSOL SUBJECT RESTRAINT

PRESENTATION

TASK 1A: **PRESENTATION EXERCISE**

CONDITIONS: Given an ASR, carrier, and duty belt.

CRITERION: Present the ASR unit from the carrier and assume positions 1–3.

Above: Inert ASR training units are available and are highly recommended for use.

Right: Training units will be clearly marked as such and should not be mixed in with live ASR units.

TRAINING GUIDELINES:

Position 1

A. The presentation of the belt-carried ASR unit should be kept as simple as possible. Many departments currently advocate that the holster/ASR unit be placed on the duty belt directly in front of the service pistol/holster. I have found that this positioning of the unit makes both training and actual accessing of the ASR faster and simpler because of the familiarity of the movements. **NOTE:** *The holster/ASR unit must not interfere with a clean presentation of the pistol!*

Regardless of where you position the holster/ASR unit, you must be able to get to it quickly and without interference. Holster flaps or strap devices should be easily opened, and the unit should be removed quickly and brought into full presentation position cleanly while using a solid grip on the device.

Your stance, always a critical component when tactically engaged, should be wide enough to provide stability yet still allow you to move quickly in any direction.

1. Presentation to Position 1

a) Remove the ASR from the carrier with the primary hand.
b) Hold the ASR back toward your body in the primary hand as shown in the illustration below and detailed in Chapter 11.
c) The support (nonspraying) arm is forward in a defensive posture to ward off blows and to keep the intended sprayee from grabbing the unit from you.
d) Complete a minimum of 10 *correct* repetitions.

2. Presentation to Position 2

a) Remove the ASR from the carrier with the primary hand.
b) Hold the ASR extended toward the sprayee, as shown in the illustration below and detailed in Chapter 11.
c) The support (nonspraying) hand may be held back toward the sprayer's body in a defensive ready position or out to the side for balance.
d) Complete a minimum of 10 *correct* repetitions.

3. Presentation to Position 3

a) Remove the ASR from the carrier with the primary hand.
b) Hold the ASR extended toward the sprayee as shown in the illustration below and detailed in Chapter 11. The index finger is placed on the top-mounted actuator.
c) The support hand is placed over the back of the primary hand for stability and support.
d) Complete a minimum of 10 *correct* repetitions.

Position 3

Position 2

TASK # 1 – AEROSOL SUBJECT RESTRAINT

STATIC APPLICATION

TASK 1B: ADMINISTER STATIC APPLICATION

CONDITIONS: Given an inert ASR, carrier, duty belt, and appropriate training target.

CRITERIA:
1. Present the ASR from the carrier and assume preferred stance position.
2. Use proper verbalization.
3. Administer an application of inert agent to the static training target, delivering the inert agent to the preferred area of the training target.

TRAINING GUIDELINES:

A. IMPORTANT! TRAINING WITH INERT OR CHEMICAL AGENT WEAPONS SHOULD ONLY BE CONDUCTED BY KNOWLEDGEABLE PERSONNEL TRAINED AND CERTIFIED IN THEIR USE. PROTECTIVE EYEWEAR MUST BE WORN!

B. The agent should be administered in short, quick bursts, each lasting from 1/2 to 1 second in duration. The eyes, nose, and mouth are the primary areas to aim for.

Different techniques may be used to ensure adequate coverage of the OA. Several short bursts may be applied using minimal movement of the canister in the primary hand to produce various sweeping spray patterns, such as those illustrated below and detailed in Chapter 11.

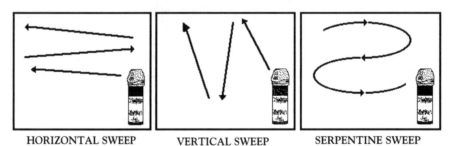

HORIZONTAL SWEEP VERTICAL SWEEP SERPENTINE SWEEP

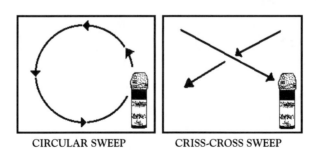

CIRCULAR SWEEP CRISS-CROSS SWEEP

Illustration 20-1. Horizontal sweep, vertical sweep, serpentine sweep, circular sweep, crisscross sweep.

C. When practicing, remember to incorporate clear, direct, and loud verbal communications (such as those given below and detailed in Chapter 11) throughout the training drill. Remember the preferred order of ask, advise, order.

a) "Police! Don't move!"

b) "Get back—don't make me spray you!"

c) "Stop!" or "Back off!"

d) "Don't rub your eyes."

e) "On the ground before you fall!"

f) "Hands behind your back."

g) "Breathe slowly."

h) "I will help you if you stop fighting."

1. Administering Static Application to Training Target

a) After presenting the ASR and assuming the preferred stance, administer an application of inert agent to the preferred area of the training target.

b) The target should be the approximate size of a human face. A standard 8 1/2 x 11-inch piece of paper with face drawn on it will do.

c) Practice hitting the preferred training target area from various distances.

Static application practice using paper target.

TASK # 1 – AEROSOL SUBJECT RESTRAINT

DYNAMIC APPLICATION

TASK 1C: ADMINISTER DYNAMIC APPLICATION

CONDITIONS: Given an inert ASR, carrier, duty belt, appropriate training target, and assistant.

CRITERIA:
1. Present the ASR from the carrier and assume preferred stance position.
2. Use proper verbalization.
3. Administer an application of inert agent to the dynamic training target, delivering the agent to the preferred area of the training target.

TRAINING GUIDELINES:

A. **IMPORTANT! TRAINING WITH INERT OR CHEMICAL AGENT WEAPONS SHOULD ONLY BE CONDUCTED BY KNOWLEDGEABLE PERSONNEL TRAINED AND CERTIFIED IN THEIR USE. PROTECTIVE EYEWEAR MUST BE WORN!**

B. The agent should be administered in short, quick bursts, each lasting from 1/2 to 1 second in duration. The eyes, nose, and mouth are the primary areas to aim for.

Different techniques may be used to ensure adequate coverage of the OA. Several short bursts may be applied using minimal movement of the canister in the primary hand to produce various sweeping spray patterns, such as those illustrated below and detailed in Chapter 11.

C. When practicing, remember to incorporate clear, direct, and loud verbal communications (such as those given below and detailed in Chapter 11) throughout the training drill. Remember to be prepared to react by moving or engaging the suspect with an alternative level of appropriate force should the situation require it.

1. Administering Dynamic Application to Training Target

a) After presenting the ASR and assuming the preferred stance, administer an application of inert agent to the preferred area of the training target.

b) The target should be the approximate size of a human face. A standard 8 1/2 x 11-inch piece of paper with face drawn on it will do. The assistant can hold this paper in front of his face while moving. Protective eyewear is mandatory to avoid serious injury.

c) Practice hitting the preferred training target area from various distances while you and your assistant take turns moving, as well as while you both move at the same time.

Dynamic application practice.

TASK # 1 – AEROSOL SUBJECT RESTRAINT

RECOVERY TO CARRIER

TASK 1D: **RECOVERY TO CARRIER EXERCISE**

CONDITIONS: Given an ASR, carrier, and duty belt.

CRITERION: Recover the ASR unit from the ready position to the carrier.

TRAINING GUIDELINES:

A. Tactically recover or reholster the unit with one hand while keeping your eyes off the gear and on the suspect. Only through repetitive practice will you be able to perform this fluidly and deliberately. The ASR should be completely secured in the carrier when possible, using any flaps or snaps incorporated into the carrier's design.

1. Recovery to Carrier

a) Recover the ASR to the carrier with the primary hand while maintaining your visual focus on the suspect/threat area. You should incorporate a scanning movement into your recovery technique, remaining alert to any movement or potential dangers in your immediate area.
b) Fully secure any flaps, snaps, or other closure devices.
c) The tactical recovery should be completed every time you present your ASR, whether in training or in the field.

TASK # 1 – AEROSOL SUBJECT RESTRAINT

TRANSITION DRILL

TASK 1E: TRANSITION DRILL

CONDITIONS: Given an ASR, carrier, duty belt, and alternative authorized use of force weapon.

CRITERION: Transition from the ASR ready position to another level-of-force option.

TRAINING GUIDELINES:

A. Transition from one level-of-force option to another must be practiced. Only by practicing can you avoid "freezing up" when faced with a serious imminent threat. When in transition from the ASR to a higher level-of-force option (impact weapon or firearm), the ASR should be immediately released (not thrown—this eats up time!) from the primary hand and the alternative weapon reached by this same hand. You must not waste the time required to recover the ASR to the carrier before accessing the higher level-of-force option when necessary.

1. Transition to the Baton

a)   The ASR is released at the same moment that the primary hand begins to execute the presentation of the baton. There is no hesitation.
b)   The baton is presented and employed in accordance with your training.

Transition to the Baton

Baton is accessed . . .

ASR is released as primary hand begins presentation of baton.

. . . and presentation of baton is completed.

2. Transition to the Pistol

a) The ASR is released at the same moment that the primary hand begins to execute the presentation of the pistol. There is no hesitation.
b) The pistol is presented in accordance with your training.
c) Cover is taken and used as appropriate or available.

Transition to the Pistol

ASR is released as . . .

. . . primary hand begins presentation of pistol.

Pistol is accessed . . .

. . . and presentation of pistol is completed.

TASK # 2 – SMOKE AND CHEMICAL AGENT WEAPON DEVICES

PERFORM SAFETY CHECK ON DEVICES

TASK 2A: PERFORM SAFETY CHECK ON DEVICES

CONDITIONS: Given an inert smoke or chemical agent weapon device and certified instructor.

CRITERIA: 1. Inspect the device and report findings to the training officer.
2. Once the findings have been reported, correct any minor defects you are authorized to correct.

TRAINING GUIDELINES:

A. Chemical agent weapon and smoke devices are commonly maintained and employed by members of civilian and military police agencies. Very often these devices are stored improperly, mishandled, or subject to neglect. This can lead to undue wear and damage to the device. It is the obligation of all officers to inspect any equipment they are issued or may be expected to employ. This is especially critical for gear such as firearms, smoke, and chemical agent weapon devices.

B. Inspection and correction of defects comprises the following.

1. Overall Condition
Inspect the device for any obvious defects:
a) Is fuze unscrewed from body? Is it damaged?
b) Is safety pin in place? Is it damaged or bent?
c) Is safety lever (spoon) damaged or bent?
d) Is the device dirty? Rusty?

2. Correction of Defects
The department member should correct the following defects if so authorized by departmental policy:
a) If the fuze is unscrewed from body, the member can tighten it.
b) If the safety pin is slightly bent, the member can fix it.
c) If the device is dirty, the member should wipe it off.

3. Turning-in of Devices
Devices are turned in if the following are found:
a) The fuze does not screw in easily.
b) The safety pin is severely bent or damaged.
c) The safety lever (spoon) is damaged or bent.
d) The device is rusty.
e) The device is corroded.
f) The expiration date has been reached.

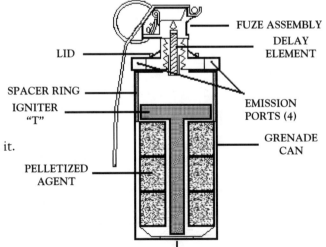

FUZE ASSEMBLY
DELAY ELEMENT
LID
SPACER RING
IGNITER "T"
EMISSION PORTS (4)
GRENADE CAN
PELLETIZED AGENT

BOTTOM EMISSION PORT (CHEMICAL CONFIGURATIONS ONLY).

Illustration 20-2. The pyrotechnic grenade discussed in Chapter

198

TASK # 2 – SMOKE AND CHEMICAL AGENT WEAPON DEVICES

IDENTIFY AND EMPLOY DEVICES

TASK 2B: IDENTIFY AND EMPLOY DEVICES

CONDITIONS: Given various types of inert devices (CAW and smoke) and a certified instructor.

CRITERIA: 1. Identify the type of device presented.
                             2. Explain in general terms when you would use it.
                             3. Demonstrate how to hold, make ready, and deliver each of them.

TRAINING GUIDELINES:

A. **IMPORTANT! TRAINING WITH CAW AND SMOKE DEVICES SHOULD BE CONDUCTED ONLY BY KNOWLEDGEABLE PERSONNEL TRAINED AND CERTIFIED IN THEIR USE.**

B. **Identify the type of device presented:** All officers should be familiar with the types of CAW and/or smoke devices maintained for use by the department. The officer in charge (OIC) or chemical agent munitions officer (CAMO) is responsible for ensuring that all pyrotechnic, aerosol, and other devices are maintained in good and serviceable order and that all authorized officers are familiar with the device's nomenclature, operational features, dangers, and employment methods. Devices should be maintained in their original packing cartons until ready for use. All devices should be clearly identified by markings on their packaging boxes/cartons/bags, as well as by markings on the device body. A color-coding identification system is also used to mark and identify the various agent-loaded devices.

1. **Smoke Grenades**
   These can be identified by written markings on the packaging materials and grenade body. Smoke grenades are generally configured with only four emission ports on the top of the canister, while chemical agent weapon grenades will have four on the top and one on the bottom.

2. **Chemical Agent Weapon Grenades**
   These can be identified by written markings on the packaging materials and grenade body. Pyrotechnic chemical agent weapon grenades are generally configured with four emission ports on the top of the canister and one on the bottom, while smoke grenades have only four emission ports on the top of the canister. Instantaneous-blast and aerosol grenades come in various configurations and must be individually identified. The type of agent loaded into the device can also often be identified by a color-code identification system:

         CN (chloroacetophenone) Color Code: Red
         CS (orthochlorobenzylmalononitrile) Color Code: Blue
         OC (oleoresin capsicum) Color Code: Orange

3. **Gas Launchers**

   These launchers come in a variety of configurations and models (see Chapter 14). They must be individually identified; and certified, professional training with them prior to use is mandatory.

C. **Explain when you would use each type of device.** This will be based on departmental policies and procedures, rules and regulations. Certified instruction should be received by all users before procurement, and MUST be received by users before individuals are issued these devices or employ them.

D. **Demonstrate how to use it.**

   All officers should have, at the minimum, a practical understanding of the different methods available that will allow them to safely deliver an authorized CAW or smoke device into an OA. The following hand-delivery techniques are provided as a reference guide for training purposes. As with any technique, thorough instruction, organized training programs, and repetition are required to achieve an acceptable level of operational competence.

1. **Preparing the CAW Grenade (as detailed in Chapter 5)**

   Hold the device firmly in the dominant hand, keeping the safety lever (or spoon) pressed into the web of the hand. Extend the dominant arm to the front of the body.

   Keeping the device extended to the front, grasp the pull-ring securely with the index finger of the nondominant hand, palm forward, thumb-side down, and simultaneously pull the pin out as you rotate your hand forward to the palm-in, thumb-side-up position. This twisting "corkscrew" motion makes removing the pin much easier. The device is now ready to be thrown.

2. **Overhand Delivery Technique**

   a) Leaving the pull-pin on the dominant hand finger, elevate the nondominant arm and extend it toward the objective area.
   b) Cock the dominant, or throwing, arm in a manner that can best be described as a combination baseball throw/shot-put technique.
   c) Piston the dominant arm forward, following the upward angle of the extended nondominant arm. Release the device at the moment the dominant arm is fully extended.
   d) Unmounted tires laid flat on the ground at varying distances from the officer can be used to develop accurate delivery ability.

3. **Underhand Delivery Technique**

   a) Holding the device in the dominant hand, extend the nondominant arm forward toward the OA, providing a visual index as well as helping to maintain balance.
   b) Bring the dominant arm to a rearward position, keeping the elbow straightened.
   c) Rotate the dominant arm forward from the shoulder. Release the device at the appropriate moment.
   d) Unmounted tires laid flat on the ground at varying distances from the officer can be used to develop accurate delivery ability.

4. **Gas Launchers and Shotgun-Delivered Agents**

   Chemical agent weapon and smoke munitions may be launched by different methods, as outlined in Chapter 14. Gas launchers are produced in a variety of different sizes, shapes, and configurations. At closer

Above: Unmounted tires can be effectively used as training targets when practicing delivery techniques.

Right: Operator practicing underhand delivery technique.

ranges, chemical agent weapons in micropulverized form can be expelled directly out of the end of the muzzle of a gas launcher and into the atmosphere (see "Muzzle Blast Dispersion Rounds" in Chapter 9).

Canister-contained agents may also be delivered to an OA when the canister itself is launched either out of the muzzle of the launcher or out of an adapter cup designed for this purpose.

While there are exceptions, the vast majority of handheld gas launchers are produced in 37, 38, or 40mm calibers.

Shotguns can also be used to deliver agents into an OA, either by firing munitions through the shotgun (as with ferret rounds) or by launching them with the use of special launching cups and launching rounds.

Because of the number of launcher designs, sighting systems, available munitions, and extreme dangers inherent when using any of these launcher systems or agents, you must seek out specific and certified training with whichever your department or agency uses.

NOTE: Accuracy in using any of these techniques will depend on your training. Dummy devices that simulate the weight and feel of the munitions you will be deploying in live action should be used while practicing.

TASK # 3 – CHEMICAL AGENT PROTECTIVE MASK

PERFORM OPERATOR'S MAINTENANCE

TASK 3A: PERFORM OPERATOR'S MAINTENANCE

CONDITIONS: Given a chemical agent protective mask, carrier, authorized accessories, and cleaning materials.

CRITERION: Perform operator's maintenance on mask.

TRAINING GUIDELINES:

A. The chemical agent protective mask must be maintained in serviceable condition at all times. Regularly scheduled, routine servicing of the mask will ensure that the equipment will not fail when needed most.

1. Inspect Mask and Carrier

a) Remove mask from carrier. Inspect all components. Immediately report any missing components to OIC.
b) Check the carrier for wear, dirt, missing straps, or broken hardware.
c) Check the mask for holes, wear, splits, rubber fatigue, dirt, or mildew.
d) Check all filter elements. Ensure that they are serviceable. Replace any filter elements that are wet, have been directly exposed to water, or have passed their expiration date.
e) Check eye lenses for scratches, cracks, or discoloration.
f) Check head harness for wear: broken or frayed straps, dirt and mildew, broken strap-securing hardware.
g) Check hood for wear, tears, holes, dirt, or mildew (on masks so equipped).

2. Clean Mask: Follow disassembly and care instructions issued with mask. If no instructions are provided, you may use the following as a general guideline:

a) Remove voicemitter-outlet cover, inlet valve caps, and protective eye-lenses or outserts.
b) Remove external filter elements. (On M17 series masks, filter elements may be left in for routine inspection and cleaning.)
c) Clean mask inside and out with a soft cloth that has been dipped in warm, soapy water. Cloth should be wrung out thoroughly before use. A soft bristled brush may also be used. **Be sure *not* to wet the filter elements!** (Filter caps should be left on until the mask is used.)
d) Rinse with a soft cloth that has been dipped in warm, clear water. This cloth should also be wrung out thoroughly before use.
e) Dry with soft, lint free cloth or allow mask to air-dry.
f) Reassemble mask.

NOTE: Many chemical agent protective masks will have a light white or blue tinted powder on the rubber components. This is a rubber preservative and should be left on.

3. Replace Filter Elements

a) Filter elements should be replaced:
— After exposure to chemical agents
— After being immersed in water
— After being damaged
— After expiration date has been reached

4. Clean Carrier

a) Empty carrier completely.
b) Brush off the carrier inside and out, removing any dirt, sand, or grit.
c) If the carrier is soiled, clean it with a brush and cold water.
d) Reassemble carrier components.

TASK # 3 – CHEMICAL AGENT PROTECTIVE MASK

DON, CLEAR, AND TEST SEAL

TASK 3B: DON, CLEAR, AND TEST SEAL

CONDITIONS: While carrying a chemical agent protective mask in its carrier, secured to your person in its normal manner and location.

CRITERION: When given a signal or upon command, remove mask from carrier, and properly don, clear, and seal mask to your face within 9 seconds.

Left: Now . . . Officers practice donning protective respirators.

Above: And then . . . The equipment has changed, but techniques are basically the same. U.S. soldiers during World War I (like those shown here) were expected to be able to access and don their masks in about 10 seconds. (Photo from the author's collection.)

TRAINING GUIDELINES:

A. In many situations, you will know well in advance if chemical agent weapons are to be employed. Masks should be donned at these times, well before the actual delivery of the CAWs to the OA. Donning speed in these instances usually will not be vital.

However, there may be times when you will need to don, clear, and seal the mask immediately when involved in an actual situation. The possibility exists that a suspect may have chemical agent weapons in his possession and may use them. The possibility also exists—especially during combined operations where several different law enforcement agencies are involved—that a law enforcement officer may employ chemical agent weapons without giving proper warning to all involved police personnel.

A situation may also suddenly change from static to dynamic, requiring immediate entry into a location contaminated with a chemical agent.

In preparation for these contingencies, you must be able to access the chemical agent protective mask, properly don it, clear it, and achieve a good seal in as short a time as possible. (See Chapter 18 for a detailed description of donning procedure.)

1. Upon Signal

a) Stop breathing.
b) Remove any headgear. (Place between legs if possible.)
c) Open carrier with support hand. With primary hand, grasp the mask by the facepiece and remove it from the carrier.
d) Grasp facepiece with both hands, thumbs inside. Open mask fully and place chin in the chin cup.
e) Pull the head harness over the head. Straighten the straps and adjust the mask quickly, smoothing the edges of the facepiece back and upward with the hands.
f) Tighten the straps in a series of twos, beginning with the bottom and working up to the temple and head straps.
g) Cover the outlet valve cover with your hand and blow out. This will clear any agent from the inside of the mask.
h) Check seal by placing hand (or hands, depending on the number of filtering canisters used) over the intake port of the canister and breathing in. If the seal is good, the mask should collapse around the face and no air should enter around the facepiece.
i) Secure the hood over head (if the mask is so equipped).

TASK # 3 – CHEMICAL AGENT PROTECTIVE MASK

WEAR MASK IN CHEMICAL ENVIRONMENT

TASK 3C: WEAR MASK IN CHEMICAL ENVIRONMENT

CONDITIONS: Given a chemical agent protective mask that has been properly donned, cleared, and sealed to the face, an approved chemical agent weapon (CN, CS, or OC), authorized contaminated training area, a certified chemical agent munitions instructor, and appropriate decontamination equipment.

CRITERION: Wear a chemical protective mask in a training environment sufficiently contaminated with a chemical agent weapon.

NOTE: ONLY CN, CS, or OC AGENTS SHOULD BE USED!

Checking integrity of the mask and seal in an environment contaminated with chemical agent weapon.

TRAINING GUIDELINES:

A. To ensure that the chemical agent protective mask is functioning properly, as well as to ensure that each individual officer has confidence in both the mask and his own ability to correctly don, clear, and seal it, the officer should wear the mask in a chemical environment under controlled conditions.

NOTE: After ensuring the proper use and functioning of the mask, it is also recommended that each officer unmask in the chemical environment and experience a limited exposure to the agent under controlled conditions. This will provide the officer with a more complete understanding of how the agents (CN, CS, or OC ONLY) work, as well as how to decontaminate himself and others afterward. The use of professionally trained and certified training officers to regulate the exposure and provide assistance as needed to those so exposed to the agent is mandatory.

Exposure time should be just enough for the effects of the agent to be experienced. One way to facilitate this is to have each officer unmask in the contaminated environment, and say in a loud voice, "My name is John Doe, my Social Security number is 222-22-2222, I work for the Metropolis Police Department."

This exercise should provide a sufficient exposure.

1. Procedure
a) Have a certified chemical agent munitions instructor select an approved CAW for use (CN, CS, or OC ONLY).
b) Have participating officers don, clear, and seal masks.
c) Have certified chemical agent munitions instructor activate the CAW delivery system (e.g., pyrotechnic, aerosol).
d) Have participating officers enter gas cloud for a few minutes while moving around, to ensure that masks are functioning properly.
e) At this point participating officers may unmask to experience effects of CAW.

SAFETY ALERT: Once exposed, certified chemical agent munitions instructors should watch for unusual reactions or prolonged symptoms. Personnel decontamination should be completed as soon as possible.

B. POSSIBLE AGENT EFFECTS (CN, CS, OC) AND FIRST AID

AREA AFFECTED	SYMPTOMS	FIRST AID
		GENERAL: Remove affected person from area. Remain calm, restrict activity. Major discomfort should disappear within 30 minutes.
EYES	Burning sensation, copious tearing, partial to involuntary closing	Keep eyes open, face wind. Keep hands away from eyes. Eyes may be flushed with cool water for 10 minutes. Blot tears.
SKIN	Burning or stinging sensation,	Sit and remain quiet to reduce.

	on moist areas. Blistering possible from heavy concentrations.	sweating. Expose affected areas to fresh air. Gross contamination can be handled by flushing the area with cool water for 10 minutes.
NOSE	Irritation, burning sensation. Nasal discharge.	Breathe normally. Blow nose to remove discharge.
CHEST	Irritation, burning sensation, coughing, feeling of suffocation. Tightness in chest, pain, feeling of panic.	Relax and keep calm. Provide reassurance.

IMPORTANT: ANY INDIVIDUAL SUFFERING SYMPTOMS THAT ARE NOT CHARACTERISTIC OF EXPOSURE TO THE AGENT SHOULD BE TRANSPORTED TO THE HOSPITAL. MAKE SURE THE ATTENDING STAFF KNOWS WHAT THE VICTIM WAS EXPOSED TO (HAVE MSDS!)[1]

C. PERSONAL DECONTAMINATION

CLOTHING: Wash in cold water. NO SOAP. Remove clothing from washer before final spin cycle and line dry outdoors. DO NOT DRY CLEAN; DO NOT PLACE IN DRYER. Consider using a commercial laundry.

PROTECTIVE APPARATUS: Maintain as described in TASK 9A.

PERSONAL HYGIENE: Take a 10-minute cold shower to flush agent from body. Do not scrub or use soap. Pay particular attention to area of the body where hair is located. After flushing, warm water and Ivory soap may be used. Blot with towel to dry.

TASK # 4 – INDIVIDUAL OPERATION SKILLS

DELIVER SMOKE OR CHEMICAL AGENT WEAPON TO A DESIGNATED OA IN A TACTICAL ENVIRONMENT

TASK 4: DELIVER SMOKE OR CHEMICAL AGENT WEAPON TO A DESIGNATED OA IN A TACTICAL TRAINING ENVIRONMENT

CONDITIONS: Given a tactical plan for the delivery of a chemical agent weapon, the chemical agent weapon and delivery mechanism, a designated objective area, scripted tactical scenario, and a certified chemical agent munitions instructor.

CRITERION: Deliver the proper type and amount of chemical agent weapon to the designated OA.

Delivering 12-gauge barricade-penetrating rounds with a shotgun while masked presents its own problems. Such usually simple things as attaining a sight picture require a little more work. Varying degrees of difficulty result from differently configured masks and equipment. Reality-based training allows us to discover and overcome these and other problems.

TRAINING GUIDELINES:

A. Training officers should develop a scripted training scenario that will require a controlled delivery of chemical agent weapons (CAW) to a specified OA. After completing a situation evaluation and tactical analysis of the scripted scenario, a plan should be developed regarding the selection, amount, and delivery method of CAW or smoke. The officer selected to deliver the CAW should then assemble the required equipment and deliver the CAW to the designated OA when so instructed.

1. All officers involved in the training block should be briefed on the plan to use the CAW. Any support elements who are to participate in the training (such as tactical marksmen and containment officers) must be aware of the intended direction of movement of delivery officers as well as intended methods of delivery.

2. The officers responsible for delivering the CAW to the OA will ensure that they have been given the proper type and amount of specified munitions and that all of their equipment is in satisfactory condition.

3. When using 37/38 or 40mm gas guns or shotguns with barricade-penetrating (ferret) rounds, the officer delivering the CAW will select a firing position that offers optimum cover and concealment while allowing the clearest line of fire to the OA and delivery point (e.g., door, window). If satisfactory cover is not available, the use of ballistic shields is recommended. The officer delivering the CAW to OA should also be protected by a covering officer armed with standard penetrating ammunition, as well as by a long-range tactical marksman. (Obvious care must be taken when using standard penetrating ammunition during training scenarios and should *never* be used in training if scenario entails the use of live role players within the OA.)

4. Officers should use cover and concealment when deploying a CAW by hand-delivery technique. The use of multiple ballistic shields and cover officers is highly recommended. Long-range tactical marksmen should also be used to cover the delivery team during the approach and withdrawal.

5. After the CAW has been delivered to the OA, the officer(s) making delivery should tactically withdraw from the area. Delivery officers should also be prepared to take immediate action in the event that suspects, rioters, or other parties react immediately to the introduction of the CAW during actual events.

NOTE: The proper selection, use, and quantity of appropriate CAW munitions is the responsibility of the chemical agent munitions officer. These considerations require that designated officers have proper training and experience.

TASK # 5 – UNIT OPERATION SKILLS

OPERATE IN CHEMICAL AGENT WEAPON CONTAMINATED ENVIRONMENT

TASK 5: OPERATE IN CHEMICAL AGENT WEAPON CONTAMINATED ENVIRONMENT

CONDITIONS: Given an OA, a certified chemical agent munitions instructor, a scripted tactical training scenario, and an authorized contaminated training area (contaminated with an approved chemical agent weapon (i.e., CN, CS, or OC).

CRITERION: Per scripted training scenario, operate effectively as a unit in a chemical agent weapon contaminated environment.

TRAINING GUIDELINES:

All officers who may be working together in chemical agent weapon contaminated areas during actual situations must be experienced in operating as a team under these conditions.

Training scenarios that incorporate live chemical agent weapons will allow these officers to discover any weaknesses or flaws in their equipment, tactics, or techniques.

Scripted tactical training scenarios should duplicate the actual operational environments and potential deployments as closely as possible.

NOTE

1. Material data safety sheets.

Chemical Compounds/ Tables/Toxicity

PROPERTIES OF CHEMICAL AGENTS CN AND CS
(Adopted from U.S. Army FM 3-9)

CN

Chemical symbol:	CN
Chemical agent:	Chloroacetophenone
Chemical formula:	CN, $(C_6H_5COCH_2Cl)$
Molecular weight:	154.59
State at 20°C:	Solid
Vapor density (air-1):	5.3
Liquid density (g/cc):	1.318 at 20°C. (solid)
Freezing/melting point:	54°C
Boiling point:	248°C
Vapor pressure (mm):	0.0041 at 20°C
Flash point:	High enough not to interfere with military use
Decomposition temperature (C):	Stable to boiling point
Heat of vaporization:	98°C

Odor:	Apple blossoms
Median lethal dose (mg-min/m³):	14,000
Median incapacitating dose (mg-min/m³):	80
Rate of detoxification:	Rapid
Eye and skin toxicity:	Temporary severe eye irritation; mild skin irritation
Rate of action:	Instantaneous
Physiological action:	Lacrimatory; irritates respiratory tract
Protection required:	Protective mask
Stability:	Stable
Decontamination:	Aeration in open; soda ash solution or alcoholic caustic soda in closed spaces
Means of detection:	M-nitrobenzene and alkali in white-band tube of detector kit
Use:	Training and riot control agent

CS

Chemical symbol:	CS
Chemical agent:	O-chlorobenzylmalononitrile
Chemical formula:	$ClC_6H_4CHC(CN)_2$
Molecular weight:	188.5
State at 20°C:	Colorless solid
Vapor density (air-1):	
Liquid density (g/cc):	1.04 at 20°C. (solid)
Freezing/melting point:	93–95°C

Boiling point:	310–315°C (w/decomposition)
Vapor pressure (mm):	—
Flash point:	197°C
Decomposition temperature (C):	—
Heat of vaporization:	53.6
Odor:	Pepper
Median lethal dose (mg-min/m^3):	61,000
Median incapacitating dose (mg-min/m^3):	10–20
Rate of detoxification:	Rapid (sublethal in 5–10 min.)
Eye and skin toxicity:	Highly irritating; not toxic
Rate of action:	Instantaneous
Physiological action:	Highly irritating but not toxic under most conditions*
Protection required:	Protective mask and clothing
Stability:	Stable
Decontamination:	Water, 5-percent sodium bisulfate and water rinse
Means of detection:	None
Use:	Training and riot control agent

* Refer to toxicity section that follows.

TOXICITY

Although normal exposure to CN and CS has generally proven safe and effective when administered by trained and experienced personnel, overexposure to CN or CS (or any chemical substance that affects the human body) can cause serious illness and possible death, depending upon the concentration and exposure time. This is especially true when people are exposed in confined or enclosed spaces.

Based on studies of the relative toxicity of CN and CS, the U.S. Army assigned values for use in determining both the lethal concentrations and time (LCt_{50}) and the median incapacitating doses (ICt_{50}) of these agents. Formulas for determining both the LCt_{50} and the ICt_{50} of CN and CS were developed based on these values.

Although these formulas can be used to estimate the approximate number of individual munitions that should be deployed into an enclosed space to incapacitate the average human being in that space, they are not exact for determining either the lethality or incapacitating ability of the agents. There are many variables that affect both when used in real-world applications. Examples of the variables include the concentration of the agent in the air; the duration of the exposure; the size, layout, ventilation conditions, and temperature of the affected area; the physical and psychological condition of persons exposed; and respiration rates.

As for possibly causing the unintended death of a subject exposed to CN or CS, it should be noted that there is an enormous disparity between the amount of either agent generally needed to incapacitate and the amount believed necessary to constitute a lethal dose.

In the case of CN, the calculated safety factor is estimated to be 700, meaning that it would take a dose of CN approximately 700 times greater than the incapacitating dose to produce a dose lethal in 50 percent of the population.

As for CS, the calculated safety factor can be estimated at 2,500, meaning that it would take a dose of CS approximately 2,500 times greater than the incapacitating dose to produce a lethal dosage.

Such relatively high safety factors indicate that the potential for administering a lethal dose of CN or CS under most conditions is low. Of course, all risks must be considered when administering any level of force, and an application of chemical agents is obviously in this category.

As for estimating the incapacitation amounts needed, the ICt_{50} formula can be used to estimate the number of devices needed to produce the desired results within the specified enclosed area. Again, although not exact, the ICt_{50} formula can also assist officers in deciding how long to wait after an application of CN or CS before attempting a forced extraction.

The formulas for estimating the ICt_{50} and LCt_{50} are provided below for information purposes. Before applying any doses of any chemical agents, however, it is strongly recommended that you seek out qualified and experienced trainers and complete a viable, comprehensive, certified training program in their use.

COMPUTING MEDIAN LETHAL DOSE (LCT_{50}) AND MEDIAN INCAPACITATING DOSE (ICT_{50}) OF CN AND CS[*]

Median Incapacitating Dose (ICt_{50})

CN—20mg-min/m^3
CS—10mg-min/m^3

Median Lethal Dose (LCt_{50})

CN—14,000mg-min/M^3
CS—25,000mg-min/M^3

TABLE 1

Example: OA is a space 9 x 8 x 10 feet. Device is a 517 CN flameless pyrotechnic grenade with 11 grams of agent.

Step 1: Compute the volumetric size of the space to be exposed by multiplying length by width by height.

9' x 8' x 10' = 720 ft.^{3} (cubic meter measurement)

720 ft.^{3} divided by 27 = 26.66 yds.^{3} (27 is the divisor because there are 27 cubic feet in a cubic meter)

26.66 yds.^{3} divided by 1.31 = 20. (1.31 used to convert cubic yards to cubic meters because there are 1.31 cubic yards in a cubic meter)

Step 2: Convert gram weight of agent in chosen munition to milligram (mg) weight by multiplying grams by 1,000.

Example: 11 g x 1000 = 11,000mg

Example: 20.37 M^3 x 20 mg-min/m^3 = 407.12mg (ICt_{50})

Step 3: Compute incapacitation and lethal amount for calculated volumetric size.

20.37M^3 x 14,000mg-min/m^3 = 285,180mg (LCt_{50})

* Adapted from *The Smith & Wesson Academy Chemical Munitions Instructor Handbook*.

Step 4: Compute the actual number of munitions to be delivered.
   (Since you cannot deliver only a portion of a munition, always round your answer up.)

Example: Divide the answer in step 3 by the answer in step 2.

- **Incapacitating Dose (ICt_{50}) Calculation:**

   407.12 mg divided by 11,000 = 0.04

   0.04 rounded up means 1 munition should produce the desired incapacitation

- **Lethal Dosage (LCt_{50}) Calculation:**

   285,180 mg divided by 11,000 = 25.92

   25.92 rounded up means 26 munitions could create a lethal environment.

COMPUTING INCAPACITATING AND LETHAL DOSE TIME LIMITS FOR CN AND CS*

Multipliers

Median Incapacitating Dose (ICt$_{50}$)

CN—0.00227
CS—0.00057

Median Lethal Dose (LCt$_{50}$)

CN—0.39660
CS—0.70921

TABLE 2

Example: OA is a space 9 x 8 x 10 feet. Device is a 517 CN flameless pyrotechnic grenade with 11 grams of agent.

Step 1: Compute the volumetric size of the space to be exposed by multiplying length by width by height.

9' x 8' x 10' = 720 ft.^{3} (cubic meter measurement)

Step 2: Choose appropriate multiplier value from Table 2 above and multiply by answer to step 1.

ICt$_{50}$ value is 0.00227   LCt$_{50}$ value is 0.39660

0.00227 x 720 = 1.634   0.39660 x 720 = 285.55

Step 3: Divide the answer to step 2 by gram weight of agent.

Time for incapacitating dose ICt$_{50}$ Time for lethal dose LCt$_{50}$

1.634 divided by 11 = 0.148 minutes   285.55 divided by 11 = 25.95 minutes

Step 4: Multiply answer to step 3 by 60 to convert minutes to seconds.
Divide answer to step 3 to convert minutes to hours (if desired).

Time for incapacitating dose ICt$_{50}$ Time for lethal dose LCt$_{50}$

0.148 minutes x 60 = 8.88 seconds   25.95 minutes

* Adapted from MSI Chemical Agent Instructor Program.

WHAT DOES IT ALL MEAN?

Using the two formulation tables and the example given above, what this means is that we can estimate that an application of one 517 CN flameless pyrotechnic grenade with 11 grams of agent should incapacitate anyone in our 9 x 8 x 10-foot OA within approximately 8.88 seconds.

It also means that we can reasonably estimate that we could create a lethal environment in our 9 x 8 x 10-foot OA if we administered twenty-six 517 CN flameless pyrotechnic grenades into it and kept our subject exposed to this total amount of CN agent for 25.95 minutes.

BOTTOM LINE: These tables and formulas are to be used only as guidelines for determining the type and amount of agent to be used in enclosed spaces. As with any level of force, only the minimal amount needed to subdue or incapacitate should be used. It is also critical that the officer responsible for the application of these agents be aware of the lethal potential of high concentrations of these agents in enclosed environments for extended periods.

Product and Training Sources

AERKO INTERNATIONAL
P.O. Box 23884
Ft. Lauderdale, FL 33307
Tel.: (954) 565-8475
Fax: (954) 565-8499

ALPEC TEAM, INC.
201 Rickenbacker Place
Livermore, CA 94550
(510) 606-8245
Fax: (510) 606-4279

ALSETEX, SAE
36 rue Tronchet
F-7500920
Paris, France
Tel.: +33 1 42 65 50 16 510
TX.: 280384 ALEXPLO F
Fax: +33 1 42 65 24 87

**ARMAMENT SYSTEMS
AND PROCEDURES, INC.**
Box 1794
Appleton, WI 54913
(800) 236-6243
Fax: (800) 236-8601

ARMOR HOLDINGS
(Defense Technology and Federal Laboratories)
Training Academy:
Tel: (800) 733-3832
Fax: (904) 741-9993
E-mail: training@armorholdings.com

BODYGUARD INDUSTRIES
Reliapon Police Products, Inc.
3112 Seaborg Avenue
Suite C
Ventura, CA 93003
(800) 423-0668

BRITISH AEROSPACE DEFENCE
Royal Ordnance Division
Euxton, Chorley, Lancashire
PR7 6AD, UK
Tel.: +44 1257 265511
TX.: 677495
Fax: +44 1257 242199

BROWNING SA
Police Department
Parc Industriel des Hauts Sarts
3ème Avenue, B-4400
Herstal, Belgium
Tel.: +32 41646555
TX.: 41233 FABNA B
Fax: +44 1257 242199

COMPANHIA de EXPLOSIVOS VALPARAIBA
Praia do Flamengo 200
20° Andar, Rio de Janeiro
22210, Brazil
Tel.: +55 21 205 6612
TX.: 021 22812 MANT BR

CONDOR SA
Av. Graça Aranha No 81 GR 409
CEP 20030-002 Centro
Rio de Janeiro
Brazil
Tel.: +55 21 220 3432
Fax: +55 21 532 4107

CURTIS DYNA-FOG, LTD.
P.O. Box 297
17335 US 31 North
Westfield, IN 46074-0297
Tel: (317) 896-2561
Fax: (317) 896-3788
E-mail: dynafog@iquest.net

DRÄGER SICHERHEITSTECHNIK GMBH
Revalstraße 1
23560 Lübeck
Germany
Tel: +49 451 88 20
Fax: +49 451 88 220 80
Web site: www.draeger.com

FUME FREE, INC.
P.O. Box 1680
Stuart, FL 34995-1680
Tel.: (561) 220-9414
Fax: (561) 221-4625
Web site: www.quickmask.com

GASMASKS.COM
Web site: www.gas-mask.com

GUARDIAN MANUFACTURING COMPANY
302 Conwell Ave.
Willard, OH 44890-9529
Tel: (419) 933-2711; toll free: (800) 243-7379
Fax: (419) 935-8961
Web site: www.guardian-mfg.com

GUARDIAN PROTECTIVE DEVICES, INC.
P.O. Box 133
West Berlin, NJ 08091
(856) 627-1919 / (800) 220-2010
Fax: (856) 627-7071
GSA contract gs07f9003D
Web site: www.guardpd.com

HECHLER & KOCH GmbH
Postfach 1329
D-7238 Oberndorf/Neckar
Germany
Tel.: +49 7423 791
TX.: 760313 HUKO D
Fax: +49 7423 79406

ISPRA: ISRAEL PRODUCTS RESEARCH CO., LTD.
16 Galgal Haplada Street
Industrial Zone,
IL-46130
Herzliya, Israel
Tel.: +972 525 55464
TX: 342590 ISPRA IL
Fax: +972 525 59146

JAYCOR TACTICAL SYSTEMS (JTS)
6142 Nancy Ridge Drive, Suite 101
San Diego, CA 92121
Toll free: (877) 887-3773; Tel.: (858) 638-0236
Fax: (858) 638-0781
E-mail: info@pepperball.com

J&L SELF-DEFENSE PRODUCTS
Fox Labs, Inc. Products
70 Defense Drive
Berkeley Springs, WV 25411
Toll free: 888-313-6400 (orders only)
Tel.: (304) 258-2900 (information)
Fax: (304)-258-3100

JANE'S INFORMATION GROUP
1340 Braddock Place, Suite 300
Alexandria, VA 22314-1651
Tel.: (703) 683-3700
Fax: (703) 836-0029
E-mail: info@janes.com

JMS LTD.
4914 Colley Avenue
Norfolk, Virginia 23508
P.O. Box 11525
Norfolk, VA 23517
Tel.: (804) 440-9145
Fax: (804) 4402748
(*For military protective masks and parts)

MINE SAFETY APPLIANCE (MSA)
P.O. Box 426
Pittsburgh, PA 15230
(412) 967-3000
Web site: www.msanet.com

PAINS-WESSEX (SCHERMULY) LTD
High Post
Salisbury
Wiltshire SP4 6AS, UK
Tel.: +44 1722 411611
TX.: 47486 PWSCH G
Fax: +44 1722 412121

SABER GROUP, INC.
Professional Training and Security Services
268 Main Street
PMB 138
North Reading, MA 01864
Tel.: (978) 749-3731
Fax: (978) 475-5420
Web site: www.sabergroup.com

SAFARILAND LTD INC
3120 East Mission Boulevard
P.O. Box 51478
Ontario, CA 91761
Tel.: (714) 923-7300
Fax: (714) 923-7400

SAGE INTERNATIONAL, LTD.
630 Oakland Avenue
Pontiac, MI 48342
Tel.: (810) 333-7811 Fax: (810) 333-7813

SCOTT AVIATION
309 West Crowell St.
Monroe, NC 28112
Tel.: (704) 282-8402
Fax: (704) 282-8424
Web site: www.scottaviation.com

SCOTT HEALTH AND SAFETY
Box 501
FIN-65101 Vaasa
Finland
Tel: +358 6 324 4511
Fax: +358 6 324 4591
Web site: www.scottsafety.com

SHALON-CHEMICAL INDUSTRIES, LTD.
25 Nahmani St.
P.O. Box 14234
Tel-Aviv 75794
Israel
Tel.: 972-768-11095
Fax: 972-768-11115

SMITH & WESSON ACADEMY
(and National Firearms Training Center)
299 Page Boulevard
Springfield, MA 01104
Tel.: (413) 846-6460
Fax: (413) 736-0776
www.smith-wesson.com

TA'AS ISRAEL INDUSTRIES, LTD.
P.O. Box 1044
IL-47100
Ramat Hasharon, Israel
Tel.: +972 3 542 5222
TX.: 03 3719 MISBIT IL
Fax: +972 3 540 6908

Sample Material Safety Data Sheet (MSDS)

Material Safety Data Sheet
Denlon K/Rose Group, Inc.

Emergency phone # 1-800-248-7342

Bio Shield®
revised 10-1-94

Section I

PRODUCT NAME: Irritant/Inflammatory Cleansing Solution
TRADE NAMES AND SYNONYMS: Bio Shield® T-12877, TF99, B-1000, R-1500
FORMULA: Homogeneous mixture of Herbal Extracts, Isopropyl Alcohol, Hydrogen Dioxide
INGREDIENT: TRADE SECRET: As defined in Hazard Communication Act 29 CFR 1910.1200 Para 1 (i) end Appendix D to CFR 1910.1200
OTHER DESIGNATIONS: Isopropyl (Isopropanol, $CH_2CHOHCH_3$) Hydrogen Dioxide (H_2O_2)

Section II - Hazardous Ingredients

ISOPROPYL ALCOHOL: Solvent for herbal extracts
Hazardous in its pure form
Flammable, ignites from a spark or open fire, non-flammable in a solution 40% or less
Percentage present in product: 3%

Section III - Physical Data

SOLUBILITY IN WATER: Fully dispersable in any proportions
VAPOR PRESSURE AT 20°C: 17.9mm Hg
FREEZING POINT: Freezes under -10°C
BOILING POINT: 85.4°C
APPEARANCE AND ODOR: Brown liquid with the smell of isopropyl and herbs

Section IV - Fire and Explosion Hazard Data

FLASH POINT: 354°F Open cup
EXTINGUISHING MEDIA: Dry chemical, carbon dioxide
SPECIAL FIRE FIGHTING PROCEDURE: None
UNUSUAL FIRE AND EXPLOSION HAZARD: None known

Section V - Health Hazard Data

This product has been thoroughly tested by an FDA approved laboratory and is to be used for recovery from the effects of chemical agents on the skin, and in the air. No physical effects and damage to human skin and body have been indicated, therefore unlimited amounts can be used. Because of the presence of isopropyl alcohol in the solution, the following precautions should be taken: prevent prolonged contact with the eyes and mucous membranes. If irritation occurs, immediately flush with water. Use only until the original effects of the chemical agent have subsided. After complete recovery, flush entire exposed skin area with plenty of water. In case of ingestion, drink water to dilute.

SECTION VI - Reactivity Data

This is a stable mixture under normal storage and handling conditions. It does not polymerize.

SECTION VII - Special Precautions and Spill/Leak Procedures

Do not store above 120°F
Transport solution in plastic or glass containers

SECTION VIII - Special Protection Information

None for normal use

SECTION IX - Special Precautions and Comments

Store in original containers - Protect from physical damage

Material Safety Data Sheet

Denlon K/Rose Group, Inc.

Bibliography

Applegate, Rex. *Crowd and Riot Control.* Harrisonburg, Va.: Stackpole, 1964.

Applegate, Rex. *Riot Control: Materiel and Techniques.* 2d ed. Boulder, Colo.: Paladin Press, 1981.

Archambault, Thomas J. and Gregory L. Rookwood. *Oleoresin Capsicum Aerosol Spray Instructor Certification.* Training manual. Bennington, Vt.: MSI Training Division, 1994.

Brown, Frederic J. *Chemical Warfare: A Study in Restraint.* Princeton, N.J.: Princeton University Press, 1968.

Chemical Agent Instructor Program Manual. Bennington, Vt.: MSI Mace Security International, 1996.

Chemical Munitions Instructor Course Manual. Springfield, Mass.: Smith & Wesson Academy, 1997.

Christopher, Lt. Col. George, Lt. Col. Ted Cieslak, Comdr. Randall Culpepper, Col. Edward Eitzen, and Maj. Julie Pavlin. *1998 U.S. Army Bio Warfare Handbook—Types, Risks, Precautions.* Fort Detrick, Md.: U.S. Army Medical Research Institute of Infectious Diseases.

Connelly, Owen. *Blundering to Glory: Napoleon's Military Campaigns.* Wilmington, Del.: Scholarly Resources, 1987.

Crockett, Thomas S. *Police Chemical Agents Manual.* Washington, D.C.: International Association of Chiefs of Police, Inc., Professional Standards Division: 1969.

FBI Academy Firearms Training Unit. *Oleoresin Capsicum: Training and Use.* U.S. Department of Justice, Federal Bureau of Investigation, July 1989.

Feng Gia-Fu, and Jane English, trans. *Tzu, Lao. Tao Te Ching: A New Translation.* N.Y.: Vintage Books, 1972.

Gilbert, Adrian. *World War I in Photographs.* Italy: Barnes & Noble, 2000.

Headquarters, Department of the Army. *Sniper Training and Employment.* TC 23-14, June 1989.

Headquarters, Department of the Army. *Civil Disturbances and Disasters.* FM 19-15, March 1968.

Headquarters, Departments of the Army and Air Force. *Military Chemistry and Chemical Agents.* FM 3-215, AFM 355-7, December 1963.

Headquarters, Departments of the Army and Air Force. *Military Chemistry and Chemical Compounds.* FM 3-9, AFR 355-7, October 1975.

Hersh, Seymour M. *Chemical and Biological Warfare*. Indianapolis, Ind.: Bobbs-Merrill Company, 1968.

Jacobs, Morris B. *War Gases*. N.Y.: Interscience Publishers, Inc., 1942.

Jane's Chem-Bio Handbook. Alexandria, Va.: Jane's Information Group, 1998.

Jane's Police and Security Equipment. Alexandria, Va.: Jane's Information Group, 1996/1997.

Jane's U.S. Chemical-Biological Defense Guidebook. Alexandria, Va.: Jane's Information Group, 1997/1998.

Jones, Tony L. *Specialty Police Munitions*. Boulder, Colo.: Paladin Press, 2000.

Kadlec, Lt. Col. Robert P. "Battlefield of the Future: 21st Century Warfare Issues." *Air University*.

Kleber, Brooks E. *The Chemical Warfare Service: Chemicals in Combat*. Washington, D.C.: Office of the Chief of Military History, U.S. Army, 1966.

Kozlow, Christopher. *Jane's Counter Terrorism*. Alexandria, Va.: Jane's Information Group, 1997/1998.

Lamb, Doug, *Pepper Sprays: Practical Self-Defense for Anyone, Anywhere*. Boulder, Colo.: Paladin Press, 1994.

Lefebure, Victor. *The Riddle of the Rhine: Chemical Strategy in Peace and War*. N.Y.: The Chemical Foundation, Inc., 1923.

Marshall, Gen. S.L.A. *Men against Fire*. Gloucester, Mass.: Peter Smith, 1978.

Nowicki, Ed and Roland Ouellette. *Oleoresin Capsicum Aerosol Training Manual*. Avon, Conn.: R.E.B. Security Training, Inc., 1994.

Powell, William. *The Anarchist Cookbook*. Secaucus, N.J.: Lyle Stuart, 1971.

Prentiss, Augustin M. *Chemicals in War*. New York and London: McGraw-Hill, 1937.

Siddle, Bruce K. *Sharpening the Warrior's Edge*. Milstadt, Ill.: PPCT Research Publications, 1995.

Smart, Jeffery K. *History of the Army Protective Mask*. Aberdeen Proving Ground, Md: Historical Research and Response Team.

Tzu, Sun. *The Art of War*. N.Y.: Delacorte Press, 1983.

Glossary

AC: Hydrogen cyanide.

adamsite: Diphenylaminochloroarsine (DM). Blood or vomiting agent.

aerosols: A suspension or dispersion of small particles, either solids or liquids, in a gaseous medium. Examples of common aerosols are mist, fog, and smoke.

alveoli: Microscopic air sac in the lungs where oxygen and carbon dioxide diffusion takes place through the alveolar walls.

anoxemia: Lack of adequate oxygen in the blood.

anoxia: Lack of oxygen.

anthrax: A highly infectious disease that affects humans and animals. A microscopic particle can be fatal. It is the biological "weapon of choice" because it is easy to make and stores indefinitely.

anti-dim: A grease, paste, or impregnated cloth applied to the inside of eyepieces to prevent them from misting up.

aplasia: Failure to produce cellular products from an organ or tissue, such as blood cells from bone marrow, after a toxic dose of mustard agent.

apnea: Cessation of breathing.

arsine (SA): A blood agent.

ASLET: American Society of Law Enforcement Trainers.

asphyxiation: Suffocation.

asthma: Difficulty breathing associated with bronchial obstruction precipitated by respiratory inhalants, toxins, or allergies. Inhaled chemical agents may cause bronchial spasms or mucous membrane swelling, producing asthma.

atropine: An anticholinergic used as an antidote for nerve agents.

biological agents: Categorized into two basic groups. The first is made up of living pathogenic microorganisms, which are mainly live organic germs such as anthrax (*Bacillus anthrax*). The second consists of toxins, which are the by-products of living organisms.

blepharospasm: A twitching or spasmodic contraction of the orbicular oculi muscle around the eye.

blister agent: Vesicants; chemical warfare agent that irritates and damages the skin, mucous membranes, eyes, and, when inhaled, the respiratory tract. Significant exposure will result in death between the second day and the fourth week.

blood agent: Cyanogens; chemical warfare agent that is inhaled and absorbed by the blood. The blood then carries the agent throughout the body where it interferes with the tissue oxygenation process.

buffering agent: Material added to prevent chemical agent from caking and to aid in dispersal in a blast/expulsion munition.

BW: Biological weapons.

BWC: Biological weapons convention.

BZ: 3-quinuclidinylbenzilate; incapacitating agent.

CA: Bromobenzylcyanide; riot control agent.

CAM: Chemical agent monitor.

CAMI: Chemical agent munitions instructor.

CAMO: Chemical agent munitions officer.

canister: A disposable filter attached to the gas mask that cleans contaminated air, turning it into breathable air instantaneously. Usually composed of layers of pads or chemicals to mechanically block or chemically neutralize chemical or biological agents. Layman's term is *filter*.

carcinogen: Any cancer-causing substance.

CAW: Chemical agent weapon.

CBW: Chemical and biological weapons.

CEOD: Chemical explosive ordnance disposal.

CG: Phosgene.

chemical agent: Includes lethal agents, incapacitating agents, and harassing agents.

chemical agent weapon (CAW): Any ammunition or equipment that disperses any chemical substances, whether gaseous, liquid, or solid, classified as a riot control agent, that will generally not cause death or serious injury in normal field concentrations.

chemical resistant hood: Hood that goes over the head, shoulders, and upper torso of the wearer. Designed to be worn in conjunction with the gas mask to protect the wearer against chemical agents such as "liquid mustard," which can cause severe burns to the head, neck, and body.

chemical suit: A full-body chemical suit that protects the entire body from chemical burns and possible contamination of all biohazardous materials. Large enough in the back area for air tanks or backpacks.

chemical weapon (CW): Any ammunition or equipment that disperses any chemical substances, whether gaseous, liquid, or solid, classified as a toxin, lethal, or incapacitating agent, that can cause death or serious injury in normal field concentrations.

chloroacetophenone (CN): Lacrimatory agent.

choking agent: A chemical warfare agent (e.g., CG, DP, chlorine, PS, CK) that produces irritation to the eyes and upper respiratory tract and damage to the lungs, causing pulmonary edema.

ciliary spasm: A spasm of the muscles of the eyelids that is usually painful and may interfere with the functioning of the eyelid.

CK: Cyanogen chloride.

concentration: The amount of chemical agent present in a unit volume of air. Usually expressed in milligrams per cubic meter of air (mg/m^3).

contamination: The end result of an individual or area exposed to NBC agents.

CW: Chemical weapon.

CWC: Chemical Weapons Convention.

cyanide: Any cyanide such as hydrogen cyanide and cyanogen chloride.

cyanogen chloride (CK): A blood agent.

cyanogens: NATO term for blood agents.

cyclitis: Inflammation of the ciliary body of the eye.

diaphragm: See voicemitter.

diphenylaminochloroarsine (DM): Adamsite; blood or vomiting agent.

dominant hand/eye: The naturally dominant hand or eye. While a small percentage of the population is codominant (i.e., ambidextrous), most people are decidedly right- or left-handed. Most people also possess one eye that is dominant over the other. May also be referred to as "primary" hand/eye in text.

dose or ct: The concentration (C) of chemical agent in the air multiplied by the time (t) the concentration remains. Dose is usually expressed as mg-min/m^3. Dose is a combination of concentration and time. The same range of effects can generally be produced by either heavy concentration or prolonged exposure.

dyspnea: Labored breathing resulting from an increased need for oxygen or inadequate air exchange in the lungs.

EOD: Explosive ordnance disposal.

ESP communication system: A self-contained electronic speech projection device. The compact, battery-operated unit clearly amplifies and projects the wearer's voice, allowing ungarbled speech to be heard in areas with high ambient noise. (Cannot be used with all masks.)

filter: A disposable canister attached to the gas mask that "cleans" contaminated air, turning it into breathable air instantaneously. Same as a canister.

FM: Field manual.

fog oil: A smoke made from a special petroleum oil.

full-face mask: Pertains to gas masks. A full-face mask protects the wearer's eyes, face, lungs, etc., from contamination.

GA: Tabun.

G-agent: Type of nerve agent; includes tabun, sarin, and soman.

gas mask: A device that fits snugly against the face to protect the wearer from breathing NBC agents.

GB: Sarin.

GD: Soman.

glottic edema: A swelling of the larynx caused by exposure to chemical agents.

H: European term for HD (sulfur mustard).

H-agent: A vesicant.

H-agent simulate: A vesicant simulant.

half-face or oronasal mask: Pertains to gas masks. A half-face mask does not protect the eyes, upper face, and forehead from biological or chemical agents. Can be used with the proper filter to protect the wearer from viruses; sometimes used as respirator when combined with separate eye protection. An oronasal mask can be an integrated component of a full-face mask, covering the nose and mouth and providing additional protection against any agent that may penetrate the seal of the face mask. The use of the oronasal mask beneath the full-face mask also reduces the possibility of the operator's exhaled breath misting up the eyepieces.

harassing agents: Riot control agents (e.g., tear gas, pepper spray, CN, CS, CR).

HAZMAT: Hazardous materials.

HC: A mixture of grained aluminum, zinc oxide, and hexachloroethane; smoke producer.

HD: Sulfur mustard; blister agent.

hexachlorethane (HC): Used to produce smokes.

HL: Mustard-lewisite mixture.

hydrogen cyanide (AC): A blood agent.

hyperventilation: Excessive rapid breathing. Results in a decrease in carbon dioxide tension and respiratory alkalosis.

hypoxemia or hypoxia: Insufficient oxygen in the circulatory system to adequately supply tissue cells.

IALEFI: International Association of Law Enforcement Firearms Instructors.

incapacitating agent: Designed to produce physiological or mental effects to incapacitate combatants without killing them. These effects may last for hours or days after first exposure to them.

inlet: The point at which air being breathed in enters the filter unit, before filtration begins. The inlet is usually a hole or series of holes or slots in the base or near the base of the filter.

inoculations: A series of injections given to humans or animals to protect them against diseases caused by biological agents.

irritant agent: A tear agent or lacrimator that affects the eyes, respiratory tract, and skin. Causes intense pain, tears (lacrimation), and sometimes nausea and vomiting.

isoamyl acetate (banana oil) test: A test used throughout the gas mask industry to detect leaks in the mask or faulty filters. If the wearer can smell a "banana" odor, the mask is either (a) not fitted properly or (b) has a used or ineffective filter.

kg: Kilogram.

L: Lewisite.

lacrimation: Secretion and discharge of tears.

lacrimator: A substance that induces the secretion of tears.

latent period: The period between exposure and onset of signs and symptoms.

lens cover: Protects the facepiece lens from scratches during handling, use, and storage.

lens outsert: Polycarbonate lens outsert snaps into place over the lens of the gas mask facepiece. Outsert provides additional impact protection. Tinted outsert helps conceal identity.

lethal chemical agents: Agents that may be used effectively in normal field concentrations to cause death.

lethality: A way of classifying CW agents as either lethal or nonlethal, but there is not always a clear distinction.

lethality time: Time in which any concentration could be lethal.

Levinstein mustard (HD or H): Blister agent.

lewisite (L): Blister agent.

lithium batteries: Powerful, disposable batteries that can remain is storage for extended periods (years) without losing their charge.

material safety data sheets (MSDS): Forms prepared by manufacturers of hazardous materials.

mechanical turbulence: The result of wind speed and surface obstacles that create eddies and gusts.

median incapacitating dose (ICt$_{50}$): The incapacitating dose of a chemical agent is usually expressed as the amount of inhaled vapor or aerosol that is sufficient to incapacitate 50 percent of exposed personnel. For example, the ICt$_{50}$ value of CN is reported as about 20 mg-min/m^3.

median lethal dose (LCt$_{50}$): The LCt$_{50}$ of a chemical agent is the concentration multiplied by the time of exposure that is lethal to 50 percent of exposed personnel. The unit used to express LCt$_{50}$ is milligram minutes per cubic meter. Using CN as an example, the LCt$_{50}$ value is estimated as about 14,000 mg-min/m^3.

MG/mg: Magnesium; milligram(s).

military mask: A gas mask that is styled for use by the military (e.g., for wear while using firearms, carrying large amounts of equipment on the body).

ml: Milliliter(s).

mm: Millimeter(s).

mm^3: Millimeter(s) cubed.

MSP: Massachusetts State Police.

mustard agents (HD): Blister agents. Can cause blindness. Physical problems may continue for 30 to 40 years after exposure.

NATO: North Atlantic Treaty Organization.

NBC: Nuclear, biological, and chemical.

NCO: Noncommissioned officer.

negative air system: Standard gas mask system in which contaminated air is pulled through a filter upon inhalation, allowing the wearer to breathe clean, filtered air.

nerve agent: G- or V-agents. Stays in the body for days or weeks. Penetrates all mucous surfaces: mouth, eyes, skin, etc. Causes a disruption of nerve impulse transmissions; in sufficient quantities may cause almost instant death.

NIOSH approved: National Institute of Occupational and Safety Hazards. A U.S. government agency that puts its "seal of approval" on items designed to protect the health and life of consumers. NIOSH has the same "Standard of Excellence" as "UL" does with electrical products.

nondominant eye: Opposite of dominant eye.

nondominant hand: Opposite of dominant hand; may also be referred to as "support" hand/arm in text.

nonpersistent agent: A chemical agent that disperses or vaporizes rapidly after release.

nose cup: Integral component of some gas masks used to reduce the possibility of lens fogging under low-temperature and high-humidity conditions.

nuclear: Pertaining to atomic in nature. Radionuclides that cause burns, cancer, and eventual death.

OA: Objective area.

OIC: Officer in charge.

oleoresin capsaicinoids: The active ingredients in capsicum. The capsaicinoid responsible for providing the pungency or hotness of a pepper (and the OC inflammatory agent derived from it) is called *capsaicin*. The percentage and quality of the total capsaicinoids and capsaicin determine the pungency and effectiveness of the OC agent.

oleoresin capsicum (OC): An oil taken from the placenta near the stem of a pepper. This oil,

233

which is an inflammatory agent, is the active ingredient used in most OC self-defense sprays. OC percentages measure only the amount of red pepper contained in the defense spray, not the pungency or effectiveness of the product.

ophthalmic: Pertaining to the eye.

oronasal mask: See half-face mask.

orthochlorobenzalmalononitrile (CS): Irritant.

outlet valve: Outlet valves are produced in a variety of designs; their purpose is to let exhaled air out of the mask without having to force it through the filter. The valve closes after use, making the respirator gastight. The force of outgoing air prevents any agent from entering when the valve is open.

outpost #3: Generally a place far, far away, often located in the most undesirable or outermost border of a department's or agency's jurisdiction.

particle size: The size of the very small particles of chemical agents in the cloud are measured in terms of microns (M). A micron is 1/25,000 of an inch, and those particles smaller than 1 micron in diameter are referred to as submicron particles. As a rule, agent dissemination by burning produces a much smaller particle than that produced by blast dispersal.

penetrant: A chemical substance designed to penetrate a protective mask or clothing.

persistence: An expression of the duration of effectiveness of a chemical agent. Persistence is of particular interest in relation to riot control agents in the matter of contamination of buildings and vehicles.

persistent agent: A chemical agent that continues to present a hazard for a considerable time after delivery.

phenyldichloroarsine (PD): Blister agent.

POA: Primary objective area.

point-shooting: A combat-shooting method that relies on the body's instinctive ability to point at nearby objects with reasonable accuracy. When a handgun is gripped properly and aligned with the arm, this innate ability to point accurately can be adapted to aim the gun quickly at close-range targets without using the gun's sights.

Porton Down: The Protection and Life Sciences Division of Porton Down, United Kingdom, is one of the premier research establishments involved in the development of protective materials for NBC environments.

positive air system: A breathing system that uses blowers or pressure to lightly blow fresh air across the face of the wearer.

primary hand/eye: See dominant hand/eye.

PS: Chloropicrin; irritant.

pulmonary edema: Swelling of the cells of the lungs associated with an outpouring of fluid from the capillaries into the pulmonary spaces, producing severe shortness of breath.

purified terephthalic acid: Used to produce Saf-Smoke.

respirator: Any device that aids breathing, includes those that protect against dust or NBC agents and machines used in hospitals that artificially force a person to breathe when he is otherwise physically unable to do so.

reverse face seal: A cushion of rubber or leather that runs around the lip of a respirator facepiece where it makes contact with the wearer's face. In some cases it is molded as part of the facepiece, sometimes manufactured separately and then attached only by the outer edge to the mask; hence its being reversed. The seal then has the appearance that the facepiece lip has been folded back inside the mask.

rhinitis: Inflammation of the nasal mucosa.

ricin: A toxin derived from the castor bean that can be spread in aerosol or liquid form. It causes blood poisoning and leads to the collapse of the circulatory system and a slow death.

riot control agents: Harassing agents—e.g., tear gas, pepper spray, CN, CS, CR. Not considered by the U.S. government to be CW agents because they are nonlethal in all but the highest concentrations.

RL: Release line.

safety factor: The ratio between the lethal and the incapacitating dose of a particular agent (LCt_{50}/ICt_{50}). Using CN as an example, the LCt_{50} 14,000/ICt_{50} 20 results in a safety factor of 700. In other words, it would take 700 times the incapacitating dose to produce a lethal dose in 50 percent of the exposed population.

sarin: One of the most common nerve agents. Sarin was the agent released by terrorists into a subway system in Tokyo in 1995.

SCBA: Self-contained breathing apparatus.

shelf life: The time a gas mask or filter will last while in storage at normal temperatures and humidity.

smoke ammunition: Any ammunition or equipment that produces obscuring smoke intended for use as a signaling tool or as a means of providing concealment for movement of personnel.

sodium bicarbonate: Baking soda.

sodium hypochlorite: Bleach, a source of chlorine.

soman: A nerve agent.

spectacle kit: Gas mask accessory for use by individuals who must wear corrective lenses.

sternutator: Sternutatory agent that induces coughing and sneezing.

stockinet/stockinette: Woven fabric applied to the external surface of a rubber facepiece to protect it from general wear and tear and chemical agents. Generally inefficient because it tends to soak liquid agent up, making removal harder, whereas plain rubber can be easily swabbed clean and decontaminated. Phased out during World War II.

STOP Team: Special Tactics and Operations Team, Massachusetts State Police, Maine State Police.

sugar: Primary fuel in a pyrotechnic munition.

support hand/arm: See nondominant hand/eye.

SWAT: Special weapons and tactics.

systemic poison: A poison that affects the whole body.

tabun: A nerve agent; discovered by the German scientist, Gebhardt Schrader, in 1936 while researching ways to improve pesticides. Sarin and Soman followed shortly after.

temperature gradient: The stability of the surface air layer is governed by the variation in air temperature within 6 feet of the ground level.

thermal turbulence: An effect produced by the sharp vertical rise of air from heated surfaces.

Tissot Principle: First used in the French Tissot mask of 1917, the Tissot Principle employs directed air flow through ducts usually molded as part of the facepiece. The purpose being that air being inhaled is channeled across the inner surfaces of the eyepieces, reducing the chances of their misting up.

toxic: Of, affected by, or caused by a toxin; poisonous.

toxicity: A measure of the quantity of a substance required to achieve a given effect.

toxins: Effectively natural poisons produced by living organisms. These poisons may be produced from animal or vegetable cells or synthesized in the laboratory. When these toxins are inhaled, swallowed, or injected into human beings or animals, they cause illness or death.

V-agent: A type of nerve agent; includes VE, VM, and VX. Discovered by British scientists in 1948.

vesicant: Skin burning effect; blister agent.

voicemitter/diaphragm: A respirator component that allows the wearer's voice to be heard while he is masked. It is usually constructed of a thin rubber membrane or cellophane diaphragm covered by a perforated disc.

wind speed: As a general rule, winds from 5 to 7 miles per hour are the most effective for the use of riot control agents. When winds are higher, the munitions must be placed either closer together or farther from the target area to ensure uniform area coverage.

wind turbulence: Refers to short gusts and lulls of wind that are variable in direction, strength, and duration. Turbulence is irregular air movement. There are two kinds of turbulence: mechanical and thermal.

WMD: Weapons of mass destruction.

About the Author

Mike Conti has been a member of the Massachusetts State Police (MSP) since 1986. He has spent the majority of his career in the uniformed division. Other assignments have included work in undercover narcotic investigations, high-crime-area community policing, and special security details. He also served for four years as a member of the Special Tactical Operations Team and was working as a death investigator when tasked by the superintendent of the MSP, Colonel John DiFava, to organize, staff, and train the new Firearms Training Unit (FTU) for the department.

During the creation of the FTU, Mike also developed a completely new firearms training program specifically geared to preparing police officers for the realities of a lethal-force encounter. The successful program, dubbed "the New Paradigm," has received nationwide attention and been profiled by the Law Enforcement Television Network (LETN). The New Paradigm was also the subject of a series of articles written by Mike for his popular "In the Line of Fire" column in *Guns & Ammo* magazine.

In addition to his *Guns & Ammo* column, Mike has produced more than 100 articles and is the author of the 1997 Paladin Press book *In the Line of Fire*. A proud member of both IALEFI and ASLET, Mike holds numerous instructor certifications relating to the training and use of firearms, chemical agents, and edged and impact weapons. He continues to work as a consultant to LETN and has appeared on the History Channel's television series *Tales of the Gun*.

Mike currently serves as the director of the MSP Firearms Training Unit and Saber Group, Inc., a private training and consulting company he founded in 1997.

Author Mike Conti. (Photo by Marcie Golden.)